Animal Husbandry Regaine

The farming of animals for meat and milk confronts a stark dilemma. While world demand from a growing and more affluent human population is increasing rapidly, there are strong counter-arguments that we should eat less meat and pay more attention to environmental protection, animal welfare and human health and well-being.

The aim of this book is to identify and explain the causes and contributors to current problems in animal husbandry, especially those related to 'factory farming', and advance arguments that may contribute to its successful reorientation. Husbandry is considered in its broadest sense, namely the productive and sustainable use of the land for the good of all (plants, humans and other animals).

Illustrated by examples, the first part of the book outlines principles and arguments necessary to engage with current problems: depletion of natural resources and destruction of environment, animal welfare, food and health, fair trade and sharing resources. The second part presents a series of constructive proposals for change and development in animal husbandry, both in the developed world and subsistence agriculture. These include more integrated crop and livestock farming systems, the ethics of animal welfare and environmental management, and the evolution of a new social contract whereby a balance is struck between the right to a healthy environment and good, safe food and the responsibility to preserve these things.

John Webster is Professor Emeritus at the University of Bristol (UK) and a former member of the Animal Health and Welfare Panel of the European Food Safety Agency. As Professor of Animal Husbandry at the University of Bristol Veterinary School, he established a unit for the study of animal welfare and behaviour, which is now the largest such group in the world. He is author of *Animal Welfare: A Cool Eye toward Eden* (1993) and *Animal Welfare: Limping towards Eden* (2005), both Blackwell. He is also co-editor of *The Meat Crisis* (Earthscan, 2010).

'[John Webster] is someone who has always used his in-depth expertise and forensic analytical skills to build a bigger picture – indeed, a comprehensive and internally consistent worldview. And the three central elements in that worldview (efficiency in the use of resources; humanity in the management of farm animals; sustainability in the stewardship of the living environment) provide the anchor points for the abundance of "specific issues" that John addresses in this text.'

– Jonathon Porritt, excerpt from the *Animal Husbandry Regained* foreword

'Both challenging and inspiring, *Animal Husbandry Regained* is surely John Webster's magnum opus. Here he makes a convincing and lucid case for placing compassion – for people, animals and the environment – at the heart of global food and farming policy and practice.'

– Joyce D'Silva, Ambassador for Compassion in World Farming, UK

'A timely and comprehensive book illustrating the central value of animals to agricultural systems and landscapes. Brim full of authoritative analysis and data that tie together the critical issues of justice, efficiency and sustainability in animal husbandry.'

– Professor Jules Pretty, University of Essex, UK and author of *Agri-Culture* (2002) and *The Earth Only Endures* (2007)

'After decades of neglect, the educated public has turned its attention to many problematic aspects of the production of food, particularly food of animal origin. In this scholarly and stimulating book, John Webster treats the reader to the best discussions of animal husbandry, land stewardship, agricultural industrialization and related issues I have ever read, from a seamless perspective of science and ethics.'

– Bernard E. Rollin, University Distinguished Professor, Professor of Philosophy, of Animal Sciences and of Biomedical Sciences and University Bioethicist, Colorado State University, USA

'John Webster belongs to a rare breed now desperately needed in agriculture – a hands-on scientist and vet who is also a broad thinker; versed in moral philosophy and in the philosophy of science; perceiving, therefore, that good husbandry like all human action must be rooted in compassion; that animals of the kind we keep on farms are sentient, conscious beings, demanding respect; and that rigorous science is vital and yet is limited and cannot be the ultimate arbiter of best practice. In this book he summarizes a lifetime of experience and contemplation. Immensely valuable.'

– Colin Tudge is co-founder of the Campaign for Real Farming, www.campaignforrealfarming.org, and author of *Good Food for Everyone Forever*

Animal Husbandry Regained

The place of farm animals in sustainable agriculture

John Webster

This first edition published 2013
by Routledge
2 Park Square, Milton Park, Abingdon, Oxon, OX14 4RN

Simultaneously published in the USA and Canada
by Routledge
711 Third Avenue, New York, NY 10017

Routledge is an imprint of the Taylor & Francis Group, an informa business

British Library Cataloguing in Publication Data
A catalogue record for this book is available from the British Library

Library of Congress Cataloging-in-Publication Data
Webster, John, 1938–
Animal husbandry regained : the place of farm animals in sustainable agriculture / A.J.F. Webster. – 1st ed.
p. cm.
Includes bibliographical references and index.
1. Livestock. 2. Food animals. 3. Sustainable agriculture. 4. Animal welfare. I. Title.
SF84.3.W43 2013
636–dc23 2012015683

ISBN13: 978-1-84971-420-4 (hbk)
ISBN13: 978-1-84971-421-1 (pbk)
ISBN13: 978-0-203-09422-8 (ebk)

Typeset in Bembo
by HWA Text and Data Management, London

Printed and bound in Great Britain by
TJ International Ltd, Padstow, Cornwall

To all the animals of the future,
especially Poppy, Jack, Tom, Griff, Felix and Mimi.

Contents

Figures

Tables

Boxes

Foreword

Jonathon Porritt

I've never understood how people can talk ever so eloquently about 'sustainable agriculture' without so much as a passing reference to animal welfare. That may have something to do with the fact that the animal welfare movement (let alone the animal rights movement!) has always operated in a separate zone from the 'green movement' – albeit with plenty of individual and organisational overlap.

Hence my personal enthusiasm for organisations like Compassion in World Farming (of which I've been a Patron for many years) and for all those who stay true to a more holistic vision of sustainable agriculture – and indeed, of sustainable development in general.

That's why I have also been such an admirer of John Webster's work. He is someone who has always used his in-depth expertise and forensic analytical skills to build a bigger picture – indeed, a comprehensive and internally consistent worldview. And the three central elements in that worldview (efficiency in the use of resources; humanity in the management of farm animals; sustainability in the stewardship of the living environment) provide the anchor points for the abundance of 'specific issues' that John addresses in this text.

To describe this as a 'balanced approach' does it a disservice. It *is* balanced – from the perspective of reasonably well-informed, pragmatic omnivores. Meat-eating, for John, is 'a fact of life', based on a 'cast-iron case for including some meat in our diets', putting it beyond the scope of agonised moralising. But he is respectful of those who don't share that view, and as incisively critical of the way meat is produced and consumed in society today as any organisation or individual whose concerns about meat-eating are principally moral rather than pragmatic.

But that kind of balance doesn't get in the way of robust assertiveness, playful irony, and the occasional slaughtering of sacred cows (including some of my own!) that is bound to get up some people's noses. For me, that's good: persuading people to build on their own special interests by embracing as big and challenging a concept as 'planet husbandry' is a hard ask.

At the heart of 'planet husbandry', as interpreted by John, lie two big and compelling ideas; justice and empathy. Set against what pretty much everybody now agrees to be 'an inherently unsustainable' projected increase in demand for food from animals, justice is less about non-negotiable absolutist positions and more about 'the good of all humans, farm animals and the living environment'.

But the biggest challenge to all of us is engaging our hearts as well as our intellects in these complex and challenging areas. Good animal husbandry can't just be about caring *for* animals; it has to be about caring *about* animals. Our 'failure to understand what it is to be a pig', to develop a properly empathetic understanding of farm animals' physiology and behaviour, remains a massive barrier to achieving the kind of 'best practice' standards of animal husbandry that John maps out in Chapter 6.

And if you want the light-hearted approach to that, watch out for the Pigs' Eye-View of Human Behaviour!

September 2012

Jonathon Porritt is the Founder and Director of Forum for the Future (www.forumforthefuture.org).

Preface

William Golding, author of *The Lord of the Flies*, once lamented that he had become best known as an author of set books for students reading English at A level (university entrance level), when his intention had only ever been to provide a good read. The stated aim of Earthscan is to 'publish books and journals on climate change, sustainable development and environmental technology for academic, professional and general readers'. This is a splendid aim and I am proud to join the ranks of its contributors. There is a problem however in targeting a book at all three interest groups. The intelligent general reader needs a comprehensive introduction to the subject that presents the main issues, and conflicts and compromises between the main issues, with clarity and enough detail so that, if they wish to dig further, they know where to dig. Professionals need sufficient knowledge and understanding to enable them to do their specialist jobs with professional competence. Academics expect every conclusion and assertion to be supported by comprehensive references to what is (immodestly) called 'the literature'.

This book considers the present and future role of farm animals in sustainable agriculture or 'planet husbandry'. It considers animal production, health and welfare, the composition and safety of food from animals, the science of sustainable development and climate change, the philosophy, politics and economics of rearing farm animals to provide us with food and other benefits. Each of these issues is examined according to the three principles of efficiency, humanity and stewardship.

This is a massive subject. I recognize that many of you within the academic and professional categories will find that a book of this length, written by a single author, is on occasions superficial or fails to give due emphasis to a point that you consider especially important. On the other hand, the general reader may find some sections, e.g. Chapter 2 'Audits of animals in agriculture' to be hard going. I write as a biological and veterinary scientist who will in 2013 have been in the business of animal husbandry for 50 years and every year of my experience has found itself into the book somewhere. I try however not to write like an academic. I am particularly fond of the words of Eric Hoffer, who wrote 'It is a vice of the scientific method that it fosters hemming and hawing and a scrupulousness that easily degenerates into obscurity and meaninglessness. In products of the human mind, simplicity marks the end of a process of refining, while complexity marks a primitive stage.' In the words of another of my heroes, Albert Einstein, I have tried to keep things as simple as possible – but no simpler. There is a short list of references and suggestions for further reading at the end of each chapter. They make no attempt to be comprehensive. Established facts are accepted without acknowledgement. Many of the references relate to occasions where I quote numbers, which you may care to check. Others relate to topics that are new or controversial. Here I do no more than cite one or two key reviews or original communications that you can use as a point of entry for a literature search using Google Scholar or some such instant repository of knowledge.

So dear readers, be you academics, professionals or real people, I hope you find some things of interest and of use from this book. Above all, I hope you find it a good read.

Acknowledgements

Special thanks to Jonathon Porritt, co-founder of Forum for the Future, for contributing the Foreword to this book and to Martha d'Andrade, Portalegre, Portugal, for the splendid picture of her Mertolenga cattle in the cork oak forests that provides the cover.

Part I

Engaging with the problems

1 Whatever happened to husbandry?

> Husbandry was once a sacred art; but it is pursued with irreverent haste and heedlessness.
>
> (Henry Thoreau, *Walden*)

Husbandry is a good word. When applied to any aspect of agriculture, by which I mean the management of land and life on the land, it comfortably embraces the facts and principles of science and economic production but it enriches these things with three special human qualities: duty, care and conservation. This opening sentence has, I confess, been stolen from my first ever book, whose relatively modest aim was to address the husbandry, health and welfare of young calves (Webster 1984). My present aim is rather more ambitious. It is to explore elements of the past, present and future husbandry of all farmed animals, viewed in a very broad perspective as the net sum of interactions between man, farm animals and the living environment wherever husbandry is practised, or abused.

The strategies that I shall present for the future husbandry of farm animals are based on the principle of respect: respect for science, economics and efficiency, respect for the welfare of the animals and respect for the living environment. I shall examine the proximate causes and contributors to current problems in animal husbandry, then advance arguments that may contribute to its recovery. Clearly it is not possible to consider the husbandry of farm animals in isolation. Due recognition will be given to the rising, but ultimately unsustainable, increase in demand for food from animals, and to socioeconomic issues such as food security and fair trade. However these issues have been well covered by other authors (FAO 2006; Wirsenius *et al.* 2010; D'Silva and Webster 2010). My focus is to seek constructive approaches to improving our impact on farm animals and their impact on the environment: simply expressed, how to do it properly. Part I, 'Engaging with the problems', will outline principles and arguments necessary to deal with current problems rather than simply grumble about them. It will consider how to reconcile our immediate needs to ensure economic and efficient production, farm animal welfare, food safety and human health, with our longer term responsibility to manage natural resources and the living environment (including coexistence between domestic animals and wildlife). Part II, 'Embarking on solutions', will present and discuss a series of constructive proposals that, in the words of my subtitle, suggest a future, or futures, for animal farming. These proposals will be considered on a global basis and will range from the most extensive pastoral systems to the mass production of food from animals in large, intensive peri-urban establishments, both assessed according to first principles of respect for science and respect for life.

Rules of engagement

The practice of agriculture: the working of the land and life on the land to provide us with the essentials of life, food and clothing (and ideally rather more than these) is, ever was and ever shall be bound by the fundamental laws of physics and chemistry – and these things don't change. This point is so obvious that it should not really need to be said. It is however frequently ignored in the planning, accounting and audit of an agricultural industry driven by short-term objectives (day-to-day competition in costs and prices) and further distorted by market speculation in futures and derivatives.

All agriculture is driven by the inexhaustible (for our practical purposes) energy of the sun. However, the capacity of the land to sustain life, grow crops and feed humans and other animals is constrained by limits to the availability and quality of the land and water. For most of recorded time, farmers created wealth mostly from land that they didn't own, at a rate that was determined by the food they could grow close to home and the feed their animals could harvest from further afield. All these things were, of course, critically dependent on a reliable source of soil and sunshine, warmth and water. The main reason why industrial agriculture in the developed world is so much more productive than subsistence farming, measured in terms of output, is because most of the inputs come from outside the farm. Materials are bought in as cheaply as possible and most of the energy comes not from the sun but from fossil fuels. I shall consider these issues in some depth. For the moment, the point I wish to make is that any audit of the long-term implications of agricultural practices, in this case the practice of animal husbandry, must take into account all the major factors that contribute to efficiency and sustainability in the use of renewable and non-renewable resources of energy, to balance in the use of finite but recyclable resources, such as carbon, nitrogen and other minerals, and to wisdom in the dispersal of agricultural wastes, which range from pig slurry to food past its sell-by date. These things are reviewed in Chapter 2, 'Audits of animal husbandry', which also explores ways in which farm animals can most efficiently and sustainably be incorporated into integrated systems for land use such as agroforestry, carbon sequestration and conservation grazing.

Chapter 3 addresses the issue of farm animal health and welfare. Welfare is viewed from two angles: 'How is it for them?' and 'What is it to us?' Their welfare can be defined by the extent to which they can meet their own physiological and behavioural needs, be 'healthy and happy'. Our concern is, or should be, an expression of the same thing. We want them to be healthy and happy. Those in direct contact with farm animals have an obvious responsibility to care *for* them: to promote good welfare through good husbandry. It may be less self-evident but this responsibility extends to us all. It is not sufficient simply to care *about* animals, however sincere this feeling may be. In the case of the food animals our responsibility to care for them requires us to support farming systems that can demonstrate good husbandry leading to high standards of farm animal welfare. At all stages of the food chain, from farm to fork, animal welfare, together with other critical issues such as stewardship of the natural environment, needs to be built into the concept of food quality and rewarded appropriately.

Chapter 4 looks into the product: what we get from farm animals. Citizens of urban environments in the developed world equate this to food, meat, milk and eggs, arranged on supermarket shelves in a form that looks attractive, hygienic and as remote as possible from its source, the living sentient animal. For traditional and pastoralist farmers animals were and still are a source of transport and draught power, fuel and fertiliser, food and clothing. Wherever possible, food and clothing is harvested without killing the animal; eggs, milk or, in the case of the Masai, blood. Killing the animal for meat is a last resort (for both parties); not something to be considered lightly but reserved for special occasions such as the return of a prodigal son. In these societies it the custom to keep as many animals as possible, possibly as a status symbol but also as a reserve of wealth to be drawn on in times of climatic or financial stress. To a Western farmer accustomed to viewing animal production simply in terms of efficient conversion of the feed I buy to the food I sell, this may appear crazy, and indeed it can lead to real problems of overgrazing and desertification. However, on the basis of a more complete audit that takes into account, to give but one example, the fact that the cattle, horses, camels or yaks are not only providing food but also harvesting fuel and fertiliser at zero cost and contributing to the work of the farm, it begins to make a lot more sense. This is a nicely exotic illustration of the danger of arguing from limited premises.

While I shall consider several ways in which farm animals may contribute to our own welfare, most of this book will inevitably focus on food production. I begin with four facts of life.

- For urban citizens in the developed world, food has become cheaper and cheaper. Average expenditure on food (consumed in the home) in the UK was 26 per cent of household income in 1966; in 2010 it was 10 per cent (DEFRA 2011).
- Notwithstanding the fact that vegetarians continue to thrive both individually and collectively, global consumption of meat consumption continues to rise because the great majority of people enjoy it and more and more can afford it. Table 1.1 presents past and projected trends in meat and milk consumption in developed and developing countries, expressed as total consumption *per capita* and consumption *in toto* (FAO 2006). In the developed world the increase in meat and milk consumption from 2002 to 2030 is predicted at 5 per cent. In the developing world (which

Table 1.1 Past and projected trends in meat and milk consumption in developed and developing countries (from FAO 2006)

	Developed countries			*Developing countries*		
Food demand	*1980*	*2002*	*2030*	*1980*	*2002*	*2030*
Meat consumption (kg/person/year)	73	78	79	14	28	37
Milk consumption (kg/person/year)	195	202	209	34	46	66
Total meat consumption (tonnes × 10^6/y)	86	102	121	47	137	252
Total milk consumption (tonnes × 10^6/y)	228	265	284	114	222	452

includes India and China) the predicted increase is 96 per cent. Even so, consumption *per capita* in 2030 would still be less than 40 per cent that of omnivores in the developed affluent north and west.

- Nearly all of us in the developed affluent north and west who eat meat and other animal products eat much more than we need to meet our nutrient requirements (e.g. for essential amino acids, minerals and vitamins). In too many cases, our rate of consumption carries a serious health risk.
- The rate of increase in world consumption of meat and milk *must* decrease and soon. It has been estimated that 'By 2030, if China's people are consuming at the same rate as Americans they will eat 2/3rds of the entire global harvest and burn 100M barrels of oil a day, or 125% of current world output' (Brown 1995).

The FAO publication *Livestock's Long Shadow* (2006) presents a comprehensive quantitative analysis of current and projected systems of livestock production and their impact on the environment, and the news is nearly all bad. I quote from one particularly apocalyptic sentence: 'the livestock sector is . . . the major driver of deforestation, as well as one of the leading drivers of land degradation, pollution, climate change . . . and facilitation of invasion by alien species'. This is, of course, all true (even the invasion by aliens bit).

Livestock's Long Shadow and the recent book *The Meat Crisis* (D'Silva and Webster 2010) expound in detail on the environmental problems arising from the expansion in livestock farming (e.g. exhaustion of resources, land degradation, global warming) and the socioeconomic problems arising from disparities between supply, affordability and demand, viewed both within communities and at a global level. These are big topics and generate many questions. However, they will be discussed only briefly (perhaps too briefly) here, partly because they have been thoroughly covered elsewhere but mainly because these accounts of excesses are peripheral to the main themes of this relatively short book, which is primarily a science and ethics based investigation of what is being done at present and how it could and should be done better. One theme that will recur when considering big issues such as the feeding of grains (food for humans) to livestock, and the impact of livestock on the environment is that most investigations, and nearly all arguments, are based on limited premises, and most controversies arise from the fact that the various protagonists operate from different terms of reference. The farmer will say (correctly) that I have to make a living in a competitive world. The shopper will say (correctly) that when two batches of food (be it broiler chicken or baked beans) appear the same, it makes sense to buy the cheaper. The nutritionist will say (correctly) that consumption of meat and animal fats carries a risk to health. The animal welfarist will say (correctly) that farm animals have the right to a life worth living and the environmentalist will say (correctly) that animal farming as currently practised in both intensive and extensive systems presents serious threats to the quality and sustainability of the biosphere. Many of my arguments in the following chapters may be criticised on grounds that I have failed to take proper account of one or more cases of special pleading. Getting my retaliation in first, I would say that everything that is relevant to the big picture should be in the big picture but should not be allowed to blot out something else. If I were to attempt to consider in detail all the inputs to the analysis of these big questions, the book would become not only unreadable but impossible to write. Thus I shall not presume to offer definitive answers to these big questions, applicable in all circumstances and in all environments. My aim is to help you think about them in a way that is comprehensive, structured, scientifically sound, economically coherent and morally just.

Chapter 5, the final chapter in Part I, steps beyond the boundaries of biology and enters the worlds of philosophy, politics and economics. To quote Erwin Schrodinger: 'The image of the world around that science provides is highly deficient. It supplies a lot of factual information and puts all our experience in magnificently coherent order but keeps terribly silent about everything close to our hearts, everything that really counts.' As a quantum physicist and Nobel Prize winner, he has the right to draw attention to the limitations of science. Chapter 5 first explores philosophy, politics and economics of animal husbandry. Classic principles of ethics such as beneficence, deontology and justice are invoked to examine the rights and responsibilities of us humans, the moral agents to our moral patients, farm animals and the living environment. Having set out a moral basis for rights, responsibilities and value judgements, I then describe how these are incorporated into law and regulations, examine their difficult relationship with parallel, amoral, economic measures of value and explore how they may be reconciled through conventional politics and 'politics by other means', the power of the people.

Husbandry lost: the rise and fall of the factory farm

I live in a small hamlet in the south-west of England that goes by the splendidly feudal name of Mudford Sock. This, and the two adjacent villages, Ashington and Limington, all appear in the Domesday Book (1086) as pasture, meadow and land fit for the plough. The records show that this land has been used for mixed farming for over 1,000 years. By any process of accounting, this must be defined as sustainable. For the last 150 years, the 'milch cow'[1] has been the main source of wealth for these farmers, with the cows obtaining as much feed as possible from within the farm itself. Simply expressed, this is good, traditional cow country, but it is not just a milk factory. Dairying forms one part of an integrated mixed farming system that can (usually) support a reasonable income and other elements of a good and honest life; fields of grain and potatoes for sale grown in fields fertilised (in part) with manure from the animals; hobby pigs reared in part for the pot but equally for the fun of competition at agricultural shows; woodlands and hedges maintained as habitat for birds and other wildlife to be admired *in vivo* or shot and eaten, according to species and taste. Over the past 1,000 years the people who work this land have suffered many crises of poverty, famine and plague but, by and large, in this corner of England the life of the land has been sustained so far and it is beautiful. Yet two of the six farms in our area have gone out of dairying in the last year. Simultaneously, in eastern England, on land best suited for intensive horticulture (too good for cows) a production unit designed eventually to house 7,000 cows under zero grazing is under construction. These small happenings, in one small corner of the world, are just small broken tiles in the shattered global mosaic. Food production from animals is rising inexorably but the life is being drained from the land.

It is necessary to understand (and sympathise with) the forces that led to the industrialisation of animal production. Within traditional, family-scale farming systems, ruminants designed primarily for meat production (sheep and goats) were expected to forage for themselves, grazing land that the family did not own and consuming food that the family could not digest. The farmer and his family would grow, cut, carry and conserve food for the house cow, since she could repay this investment through the production of milk. Her male calf would probably be killed and eaten as veal before it began to compete with its mother for food. Poultry and pigs (where culturally acceptable) acted as scavengers which ate food that would otherwise have gone to waste. Young pigs intended to provide food for the table or for preservation and sale as cured ham or bacon were kept in a sty and fed on house and farm swill to ensure they fattened as quickly and as cost-effectively as possible. Breeding sows and hens would usually be given the run of at least part of the farm, since this let them forage for some of their own food and reduced the cost and labour of housing, bedding and manure disposal. In many African villages hens may not actually be offered any food at all for months on end, yet are able to scavenge enough to produce one or two eggs per week, containing enough protein to sustain healthy growth in a child subsisting otherwise on a diet consisting largely of starch (Roothaert *et al.* 2011). The traditional farmer who allowed his pigs and hens free range on his farm was not being inherently kinder than the manager of a modern poultry unit, he was simply adopting the strategy most appropriate to his own circumstances. The key to the success of this system of food production from animals was that, so far as possible, the animals harvested their own food and spread their own manure. Nobody could accuse the African hen subsisting to a large extent on a diet of ticks and insects of consuming food that should properly have gone to feed the people: it never gets the chance! This is a first illustration of one of the main themes of this book.

> Animal farming is most efficient when the needs of the animals for feed are complementary to, rather than in competition with the food needs of humans.

These simple principles formed the basis for livestock farming from the beginning of agriculture, almost to the present day. We can (with sympathetic hearts and full bellies) commend the approach as demonstrably sustainable. However, it provided little more than subsistence for most of the farmers, most of the time. The individual units were too small and distribution too difficult. Moreover when farmers do not own the land, and have no property rights, they have no incentive to work towards increasing the value of the land. This undoubtedly contributes to problems of overgrazing and land degradation. The first agricultural revolution in recent history came about when the rich began to purchase and enclose the land (to the rage of the commoners). The property owners employed large numbers of labourers, added horse power, horse-drawn machinery and, most critically, investment in long-term strategies designed to exploit the natural wealth of the land not just for immediate return but also to accumulate wealth down the generations. The ethics of all this may be questioned but it was highly efficient as measured by Adam Smith's criteria since it achieved economies of scale and synergies of labour whereby each contributed their special skills. The workers may have been poorly paid but they did get access to money, which is more useful than being paid in turnips since it creates freedom of choice. Freedom of choice drives competition amongst the producers so the invisible hand drives the market to the satisfaction of all, since we are all consumers. The Adam Smith model of wealth creation through the unregulated operation of the 'invisible hand of the market' has long been considered a dangerous oversimplification, even within the limited context of human society. More seriously, in the present context, it has no mechanism to incorporate the immediate welfare needs of the farm animals or the long-term needs of the living environment. At the time, however, it provided a pretty comprehensive model of agricultural economics because the wealth was generated almost entirely from the land itself and was therefore sustainable.

Hernando de Soto (2000) has argued that one of the greatest barriers to progress in farming development, indeed to democracy itself, in the third world is that the people do not own the land. Without property rights there is no possibility to invest, no possibility to achieve economies of scale and no opportunity to accumulate wealth. Yet when everybody is given their own tiny bit of property (two fields and a cow) this fails to meet the requirements of the Adam Smith model for economies of scale and synergies of labour. De Soto appears to suggest that what the third world needs to kick-start the accumulation

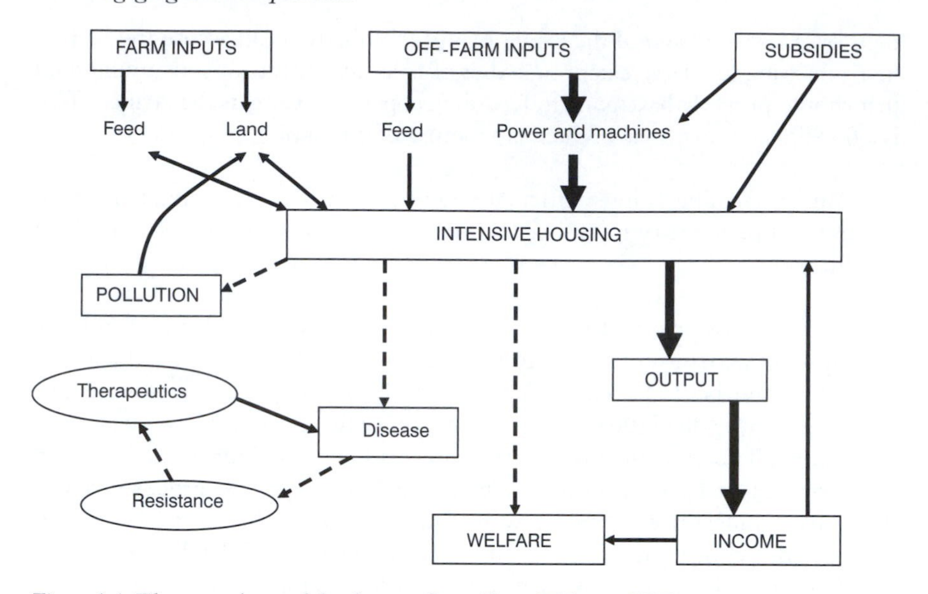

Figure 1.1 The genealogy of the factory farm (from Webster 2005)

of wealth is a benign equivalent of the Enclosures Act which gives the land to some of the people whose purpose (and whose reward) is to organise labour and resources for the benefit of all; to produce food that the people want to buy and to husband the land in a sustainable way to avoid desertification. This might seem unfair on the masses but the alternative, giving very small amounts of land to all the people, as in Zimbabwe, can lead to disaster, measured both in terms of food production and environmental sustainability.

The second industrial revolution in livestock farming really began only about 70 years ago, and only in the industrialised world. It is spreading through China and South-east Asia, but in underdeveloped countries it has hardly started. The single key that exposes the differences between the modern industrialised livestock farm and all that precedes it is that the output of the factory farm is no longer constrained by what can be produced off the land it occupies. Most or all of the inputs to the system, power, machinery (e.g. tractors) and other resources (e.g. food and fertilisers), are bought in so that output is constrained only by the amount that the producer can afford to invest in capital and other resources and the capacity of the system to process them.

Figure 1.1 outlines the genealogy of the intensive livestock farm, as typified by modern intensively housed pig and poultry units. Some feed for pigs and poultry (e.g. cereals) may be grown within the farm enterprise, but this, along with purchased feed supplements to ensure a balanced diet, is trucked onto the unit and dispensed to animals in controlled environment houses by mechanical feeding systems. The capital required to set up such a unit is high. Mechanical and electrical power is used to control temperature and ventilation in the intensive housing units, to dispense feed and to remove and disperse the manure. Factory farming was born when it became cheaper, faster and more efficient to process feed through animals using machines than to use human labour or let the animals do the work for themselves. Once the high set-up costs had been met, the input of cheap energy and other resources from off-farm was able to increase output and reduce running costs. In consequence, poultry meat from chickens and turkeys, once the food of family feasts, is now the cheapest meat on the market.

Potentially harmful outputs from intensive livestock systems (dashed lines in Figure 1.1) include increased pollution, infectious disease and abuse of animal welfare. I emphasise the word 'potential' to avoid succumbing to the simplistic mantra 'intensive bad, extensive good'. Most of the pollutants emerging from intensive livestock units are, in fact, organic fertilisers. The problem is not the material itself but too much of it in the wrong place, an inevitable consequence of divorcing the husbandry of livestock from the husbandry of the land. Bringing animals off the land and into close confinement inevitably increases the risks of infectious disease. To combat this increased risk it has been necessary to introduce strict new strategies to eliminate or at least reduce exposure to infection through improvements to biosecurity and hygiene. Where exposure to infection cannot be eliminated it is necessary to develop routine disease control measures through the use of vaccines, antibiotics and other therapeutic drugs. The key to biosecurity in an intensive pig unit is to preserve a minimal disease status in the herd by preventing any infection getting into the unit from outside. This requires strict controls on the movement of animals and stock keepers who shower and don protective clothing before entering the unit. This will normally ensure the health of the animals (one essential element of welfare) but there are obvious limits to the expression of natural behaviour in a large isolation hospital. The key element of hygiene is to minimise contact between animals and their excreta. It is important to remind ourselves that one of the main reasons for putting laying hens into battery cages was to reduce the risks to the animals and to food safety of infectious diseases such as those carried by the *Salmonella* species of bacteria.

The other essential requisite for the control of infectious diseases in intensive livestock units has been the use of veterinary preparations, therapeutic agents such as vaccines, antibiotics and antiparasitic drugs. If access to cheap power had been all that was necessary for the success of intensive livestock farming, then this industrial revolution would have occurred in the 1920s. In fact it did not really take off until the 1950s when antibiotics effective against the major endemic bacterial diseases of housed livestock became cheap and freely available. Alternative, subtler approaches to disease control, such as the development of specific vaccines and strains of animals genetically resistant to specific diseases, have also contributed to the commercial success of intensive systems, especially in the case of poultry. However, it is fair to claim that industrialised farming of pigs and poultry has, for the last 40 years, been sustained by the routine use of antibiotics, coccidiostats and other chemotherapeutics to control endemic diseases. In some cases these diseases could be life-threatening. In most cases, however, chemotherapeutics have been used routinely to increase productivity by reducing the effects of chronic, low-grade infection.

However, the era when it was possible to rear pigs and poultry under blanket cover of prophylactic antibiotics is coming to an end, in Europe at least. EU legislation to impose a complete ban on the use of antibiotics as 'zootechnical additives' or 'growth promoters' came into effect in 2006. The main reason for this ban is increasing concern that the development of microbial resistance to antibiotics used as growth promoters will pose an increasing risk to the health of the animals and, especially, to us the consumers. The scientific evidence in support of this legislation is not entirely convincing, mainly because the antibiotics currently licensed as growth promoters differ in character from those used therapeutically in veterinary and human medicine. Indeed, there is a real danger that this new prohibition will suffer a similar fate to previous worthy attempts at prohibition, and do more harm than good, partly through an increase in the black market, but more likely through a large increase in the 'legal' prescription of genuinely therapeutic antibiotics to farm animals. However, on balance, and in time, it has to be a good thing, both for the animals and ourselves, to restrict the routine use of antibiotics in livestock farming.

It is also simplistic to assume that intensification of livestock production is inherently detrimental to animal welfare. I consider this in some detail in Chapter 3. At this stage I would simply state the obvious fact that in any system where hundreds of thousands of animals are contained within a single building it is impossible to care for each animal as an individual. Any animal that falls behind the average by virtue of ill health, impaired development, reduced fertility or reluctance to compete at the feed trough, has little chance of being nursed back to normality through sympathetic stockmanship. Indeed, factory farming has imposed a neo-Darwinian variation on the principle of the survival of the fittest.

Planet husbandry

The most critical of all the facts of life is that the terrestrial elements of earth, air and water, together with the fire of the sun, are all the resources we've got. Thus we have no option but to work the land on which the sun shines, the rains fall and over which the wind blows, to provide and sustain our wants and needs. We need food. We want good food. It is also good husbandry to harvest land resources in the most appropriate and sustainable way. It is good husbandry to ranch reindeer in Lapland, even though it does not generate much income. It is not good husbandry to destroy the Amazon forest in order to grow soya beans to feed cattle, even though the short-term financial return may be compelling.

So far as we humans are concerned, we derive far more benefit from the land than cheap food for all and large profits for a few. We both want and need a satisfactory habitat that not only meets our physical requirements for food, shelter and security, but also makes us feel good. In the context of the countryside, these added benefits may be defined prosaically in terms of such things as amenities and recreation. However, recreation is another good word since it embraces a wide range of pleasures: the physical satisfaction of a walk in

a b

Figure 1.2 Alternative husbandry systems for beef cattle. a) agroforestry: Mertolenga cattle in the cork oak forests of Portugal (thanks to Martha d'Andrade); b) a beef cattle feedlot in North America (© Dave Hughes, istock)

the fresh air, the sight of a rare bird or the spiritual satisfaction of a sunset. The impact of animal farming on the quality of the countryside can range from the sublime to the grotesque (Figure 1.2).

Livestock farming becomes good husbandry only when it makes an efficient and sustainable contribution to agriculture in general. The word agriculture derives from the Latin *agros cultura,* which literally means 'cultivating fields' – which rather suggests that the task of agriculture is to subdue nature and turn it into neat green and gold squares designed exclusively for the production of food for us. If that were all, I would not be moved to its defence. We need a broader definition of agriculture defined as management of the land and all that live off the land to create resources for immediate use, ranging from food to footpaths, and lasting value measured in terms of sustained production, healthy ecosystems, viable rural communities and beauty itself. Thus this book is not just about farming and food production. It will take into account any human acts (or omissions) that may influence, either directly or indirectly:

- the exploitation of the resources of the land to produce goods and services to meet human needs and human wants;
- the quality, conservation and sustainability of the resources of the land and the plants and animals that depend on these resources.

My central aim therefore is to explore the present and future role of animal farming within the broad context of the proper lasting care of the land and all that live off the land. This may, without exaggeration, be described as 'planet husbandry' and defined as follows.

> The aim of 'planet husbandry' is to manage the land in such a way as to produce goods of value and to enhance the quality of life for humans, animals (domestic and wild) and all elements of the living environment. This aim should embrace not only our immediate needs, it must also be directed towards the sustainable management of the resources of the planet.

I repeat: this is a big subject. I am reassured however by the words of Albert Einstein who said 'Everything should be as simple as possible, but no simpler'. It is my thesis that, for most practical purposes, the multiplicity of influences on agriculture, food production and land use can be addressed according to first principles and categorised within a manageable number of proximate causes. These first principles and proximate causes are reviewed under the generic headings of resource use, food quality, environmental quality, animal health and welfare, ethics, politics and economics. Each argument is supported by evidence, although this will be illustrative rather than comprehensive, mainly so that the trees do not obscure the wood. My aim is not to subdue you with facts but to inspire you to think.

Aims, opportunities and responsibilities for planet husbandry

Aims and opportunities for total agriculture are identified in Table 1.2 as production of food and other goods for human use, management of the land as an amenity and stewardship of the environment. Every opportunity for the individual carries with it a matching responsibility. These responsibilities do not relate simply to the provision of food and other goods for humans; they extend to sentient farm animals and, indeed, to the welfare of all life on the land. Moreover they apply to us all, because we all live off the land.

Food production

Food production may be considered within three categories: food produced directly for human consumption, food produced for animal consumption and food for human consumption produced *from* animals. These categories are by no means mutually exclusive; indeed a prime aim of sustainable agriculture should be to ensure that they are made as complementary as possible. The responsibility of farmers to provide wholesome food for direct human consumption is self-evident. However the *opportunity* for farmers, not only those involved in subsistence farming but also in the developed world, to make an economic living from the production of food simply as a commodity has diminished for reasons that have little to do with productive efficiency or the quality of the product. Food for the urban rich can be produced most cheaply where the people are poor and land and labour are cheap. In developed countries where the land is expensive and people can demand a decent wage, agriculture has sought to compete by buying in resources of fertiliser and feed to augment the output of the land, machines to work the land and service the animals, and fuel to drive these machines. However, the energy and materials that they need to buy to augment the output of the land are finite and, as they become scarcer, they become more expensive.

While it is fashionable, easy (and sometimes correct) for urbanites to criticise Western 'factory' farmers for their production methods, no rational person could deny the need to produce safe, wholesome food for direct human

Table 1.2 Aims, opportunities and responsibilities for world husbandry

Aims	*Opportunities*	*Responsibilities*
Food from vegetables	Food for humans commodities value-added products	Soil quality Pollution control Preservation of habitat
	Food for animals competitive complementary	Complementarity
Food from animals	Commodities value-added products	Public health Pollution control Animal welfare
Non-food items	Fibres (cotton, wool), leather Draught power Biomass and biofuels	Animal welfare Soil quality Aesthetics
Amenities and recreation	Access to countryside Farm holidays Field sports	Health and safety Humanity, utility, stewardship
Stewardship	C & N cycling and sequestration Water management Wildlife management (flora and fauna) Enrichment and inheritance of capital	Support from society!

consumption at a reasonable price. It is possible to marshal a more cogent case against the rearing and slaughter of farm animals to provide food for humans and the growing of cereals and other crops simply to feed animals, when this food could, in theory, have been fed directly to hungry people. Evidence cited in support of this argument will include allegations of misuse of resources, abuses to animal welfare, pollution from intensive livestock units and injustice to the poor. While such concerns are real, they have, to date, had no significant impact on the international livestock industry. World consumption of foods of animal origin continues its inexorable rise, and rises in proportion to the amount of money available to spend on such food. World consumption of meat rose from 70 million tonnes in 1960 to 240 million tonnes in 2002, with the greatest increases recorded in the developing countries (Speedy 2003). Moreover, total cereal consumption *per capita* has also risen in proportion to meat consumption in consequence of the increasing proportion of cereals used as animal feed. It is a fact of life that most of those who can, will eat food from animals. Nevertheless, it is an inescapable fact of life that this rate of expansion cannot be sustained indefinitely. It has to be reduced and reduced quickly.

Non-food items

As our capital reserves of energy in the form of fossil fuels become scarcer and more expensive, the capacity of the land to generate renewable energy will become ever more important. One (extremely) conspicuous example is, of course, the construction of wind farms. Since the wind turbines do not take up much room on the ground and need to be spaced out, they can make a significant contribution to the total value that can be generated from an area of open land. Even if they do not really qualify as agriculture, they are not incompatible with land use for food production, especially meat and milk from pasture. On the other hand, it probably makes more sense to site them out at sea where the wind blows longer and stronger. An alternative approach is to harvest the energy of the sun in the form of crops grown for biomass or processed to produce biofuels. Obviously this conflicts with the use of land for food production and has generated a number of apparently powerful but conflicting arguments. This is another example of conflict arising from the fact that different protagonists argue, innocently or deliberately, from limited premises and often with little respect for numbers. The implications of growing crops for fuel on carbon and energy budgets will be discussed in more detail in Chapter 2. As a trailer for Chapter 2 I suggest that the production of biofuels as a portable source of energy for motorised vehicles will prove to be more successful than the production of biomass *per se.* This prediction is based on something that I take to be a betting certainty: namely that all developed societies will, sooner or later, come round to nuclear power as the main source of the electricity we demand to sustain the lifestyles we have come to expect. Nevertheless, I do not foresee a society propelled entirely by electric cars and nuclear aeroplanes. Monocultures of (e.g.) sugar cane or soya beans are likely to become ever more important as sources of biofuels to drive vehicles that cannot be linked to the national electricity grid (not least the tractors that work the land). However, they do present some of the greatest problems of industrialised agriculture. It would be of little lasting value if biofuels extended the shelf life of the automobile for another 30 years and then the land died of exhaustion. On the other hand, production of biofuels from (e.g.) oilseed crops generates large amounts of high protein residues that can make a significant contribution to animal feeds.

At the other, primitive, end of the spectrum of non-food items is the production of wool, cashmere and other structural fibres from sheep, goats and camelids (e.g. llamas, alpacas). Such extensive, low-input, low-output ranching has a superficial appeal to romantics and can make a sensible, if minor, contribution to good land use in certain areas, typically where the land is poor and labour is cheap. However, the reality may not match the image. Combing the fleece from an alpaca or Angora goat sounds like a good and gentle pursuit but it doesn't generate much income and animal welfare costs money. In the UK, the price of wool barely covers the cost of shearing the sheep. In Australia, Merino sheep, ranched in huge flocks almost exclusively for quality wool production, may in drought conditions starve to death because they are simply not worth feeding.

Amenities and recreation

Nearly all the land considered in this book is in private or state ownership and managed in such a way as to reward the owners for their investment in providing goods and services for the benefit of all that can afford them. This living, managed countryside is one of our most sublime and most threatened of resources. What's more, it is one we all need; we use it all the time, although familiarity tends to breed indifference. Images of the tiger in the vanishing jungle or the polar bear in the vanishing Arctic, acquired vicariously in front of the television, generate much wider concern than (e.g.) the disintegration of a community of hill sheep farmers.

What we are witnessing in the case of the tiger and the polar bear is, of course, destruction of natural habitat, an abuse of our primary responsibility to nature in the wild, which is to leave it well, alone. (The comma is significant.) It is an inescapable fact of life that man has dominion over life and land, whether we like it or not. Wherever man is present in numbers it is we who determine quality of life for the other animals and the quality of the environment that we share. We may choose to create sanctuaries for tigers but condemn hens to cages, to make a pet of the hamster but poison the rat. We may profess respect for the intrinsic value of the tiger but our actions in regard to the tiger or the hen, or indeed all life on the land, are driven by their value to us, measured in terms of the satisfaction we derive from caring about (if not caring for) the tiger or hen. If we are to meet our responsibilities to get things right, we must first acknowledge the extent to which our motivation towards nature is driven by the desire to obtain personal satisfaction from nature.

Most people in developed societies could, if we so wished, spend our entire lives within city limits and not spare a thought for the living countryside so long as it continued to provide the resources essential to our survival. However, for most plants and wild animals, the countryside is all they've got; they cannot rely on someone else to exploit it on their behalf. Thus, for both moral and practical reasons, any strategy to exploit the potential of the countryside as an amenity for human satisfaction must be matched by a responsibility to the conservation of quality in a biodiverse habitat. This acknowledges the duty of *stewardship* within the context of planet husbandry: a lasting duty that extends not only to the landowners and their families, not only to the human race, but to life itself.

Under the general heading of amenity and recreation I include field sports like hunting, shooting and fishing. In a strictly ecological (if not a moral) sense, these things are just another form of the management of animals by humans, to be set alongside the farming of domestic animals and the planned management of wildlife. The nominally impersonal aim of most wildlife management is to preserve a balance of nature within a given habitat to ensure healthy and sustainable populations of indigenous flora and flora: in essence to try and preserve nature in its natural state. In practice, this is not possible in any environment under human control. Inevitably we make choices in favour of some species over others: e.g. the red squirrel over the grey squirrel, the English over the Spanish bluebell. These two choices are justified on the basis that they favour the indigenous over the alien. This argument has some ecological merit but we cannot avoid the fact that they are human choices as to the sort of environment we want, rather than one that evolves by natural selection.

The issue of what species to cull and when to cull them presents some difficult questions of practical ethics that are considered further in Chapters 5 and 8. The major problems arise in relation to management of species at the top of the food chain that have no natural predators, other than man. With some species, especially obligate carnivores such as raptor birds (eagles, kites), population numbers are primarily regulated by the size of their prey population (e.g. small rodents). In these circumstances, provision of suitable habitat should be sufficient to achieve a stable balance of nature. With other species, such as the African elephant or the omnivorous badger, which has no natural predators in the UK, it is very difficult to avoid controlled culling as a population measure to prevent numbers expanding to the point where the balance of nature breaks down. If and when culling is to be done in the interests of sustaining the quality of the habitat for wildlife, and indeed the entire living environment, then our responsibility is to respect the principles of utility and humanity (Table 1.2). Utility implies respect for the utilitarian principle of the greatest good for the greatest number; humanity implies that any act likely to cause harm should be carried out in such a way as to minimise suffering.

The ethics of field sports is a big issue and not strictly relevant to the main themes of this book, although I have discussed it elsewhere (Webster 2005). The main argument against is, of course, the moral objection to the killing of animals for pleasure. One of the main arguments in favour is the utilitarian argument that those who derive satisfaction from the killing of animals for sport confer a value on the animals they hunt and thereby help to conserve the species and add value to the habitat over which they hunt. In the context of this book, namely husbandry of the living environment, it is only necessary to point out that, so far as the animals are concerned, the amount of suffering associated with dying and death is determined by the intensity and duration of pain, fear, malaise (etc.) involved in the dying process. The motivation of the human doing the killing, largely governed by whether the animal is classified, quite arbitrarily, as a food animal, game or vermin, is entirely irrelevant to the animal getting killed.

Investment in resources

Traditional agriculture, the family farm, is by definition, a dynastic business. Farmers see it as their duty to increase the value of their property and ensure that the land is in good heart when it passes to their descendants. This is an essential part of good husbandry. It should be self-evident that these principles should be applied on a global scale since we all have a responsibility to maintain the planet in a state fit for future use. However, human nature being what it is, the values of the family farm will only scale up to a national or international level if they are given political support. The desire of the farmer to enhance the quality of his land for the sake of his children may be morally admirable and environmentally friendly. It is also a primitive manifestation of the selfish gene; action designed to favour the family line; i.e. not so much altruism as an extended form of self-interest. This crude element of sociobiology was famously expressed by Margaret Thatcher as 'There's no such thing as Society. There are individual men and women and there are families.' Whatever may have been the unspoken subtext to her message, I do not interpret it to mean that there should not and cannot be such a thing as a good society, merely that it doesn't come naturally. It has to be created by men and women of good will with foresight and competence and protected by law.

The key to sustainability in traditional agriculture, i.e. food production from the family farm, has been to preserve and enhance the quality of the soil so that it will continue to yield bounteous crops for man and animals. It is right and proper that the pioneer movement for organic farming should be called the Soil Association (www.soilassociation.org), even though most people who buy organic food may be motivated more by a concern to preserve their own life than that of the soil (self-interest, once again!). One of the most damaging consequences of industrialisation and specialisation in agriculture has been the destruction of soil quality and the pollution of water courses, typically through the continuous production of single crops and high inputs of inorganic fertilisers. It is important to stress the obvious point that traditional agriculture could not and cannot be sustained without input from the farm animals: to work the land, to exploit the crops that we do not or cannot eat and to enrich the soil with their manure.

The concept of planet husbandry presents opportunities for investment in resources that go far beyond the conservation of the soil to sustain food production. It is in the nature of humans, world-wide, to congregate in cities, but cities are biologically crippled, depending for life-support on the countryside to provide the essentials of life (food and water), and remove and if possible recycle the waste. It is something of a cliché but nonetheless true to say that countryside and parkland are the lungs of the city. It is possible to carry this metaphor further and claim that they are the lungs, liver and kidneys. They can and should be the source of clean, fresh air, and a huge green factory for processing the outcome of our conspicuous consumption, recapturing and recycling that which is of value, and removing or sequestering that which may cause harm. Would that this were so. Industrialised agriculture conspicuously fails the test of sustaining the life of the land. It is a major contributor to pollution from excess release of nitrogen and phosphorus, with damaging consequences for waterways, wildlife and biodiversity. It is also fashionable to vilify the humble cow for belching methane to excess and so contributing to global warming. These are real and important issues, and for this very reason they need to be examined with care and a proper respect for numbers. When we consider the great themes of sustainability – energy use, carbon and nitrogen cycling, the production, recapture and sequestration of greenhouse gases – numbers are not boring, they are critical. While it should go without saying that concerns for sustainability and environmental quality are proper and essential concerns, they are too often presented in ways that highlight villains and proffer solutions on grounds that are childishly (or cynically) innumerate: arguments based on the premise that because something is harmful when present in excess it must be defined in absolute terms as 'a bad thing'. Paracelsus (1493–1541) wisely stated 'the poison is the dose'. He was talking about medicine but the same applies to agriculture. Biologically crippled factory farms may be a major source of pollutants but pollutants are just ill-distributed overdoses of valuable fertiliser. Properly managed mixed farming of crops and animals is integral to the cycles of life.

Husbandry regained: futures for animal farming

The principles of good husbandry may be defined as follows:

- efficiency in the use of resources for production of goods;
- humanity in the management of farm animals and wildlife;
- sustainability in the stewardship of the living environment.

These core principles should be seen as complementary, rather than mutually exclusive. Clearly these principles are currently being abused, and the extent of this abuse has grown enormously in the last 50 years with the expansion of industrialised agriculture (which is a minuscule period, a mere blip, in the history of agriculture). The technical and socioeconomic reasons for this period of agricultural evolution can be interpreted simply in terms of the unthinking and amoral processes of short-term economics and natural selection through survival of the fittest: that which works best or yields the best economic return in the environment of the moment is the most likely to succeed, at least until it seriously degrades the environment in which it thrives. It is undeniable that advances in agriculture, combined with advances in medicine, have been major drivers of improvements in human health and welfare. It is equally clear however that many current agricultural practices, especially those involving the farming of animals, are degrading the environment and abusing the animals on which these practices depend. We cannot go on as we are and have to explore and develop new approaches. This is an issue dictated as much by the facts of biology and ecology as by the principles of morality. Many of the decisions we make in future will have a moral dimension, but it will be an assisted morality, since resistance to change could make life very uncomfortable for us in the not too distant future.

Part II of this book, 'Embarking on solutions', explores new approaches to reconciling the aims of efficiency, profitability, sustainability and respect for quality of life in animal farming. Chapter 6, 'Better, kinder food', deals largely with methods of production. It builds on the biological and ethical principles outlined in Part I and explores routes to improvements in productivity, complementarity (Table 1.2), animal health and welfare within relatively conventional animal production systems, including intensive (high input–high output) commercial farming of animals for production of cheap food as a commodity, pastoral and extensive (low input–low output) animal farming, integrated farming systems (e.g. agroforestry) and the specialist production of added-value food from animals (e.g. organic and high welfare food). It also describes new approaches to quality assurance (QA) in relation to food from animals. Quality is measured not only in terms of the appearance, taste and safety of food but also in terms of quality of life for the food animals and quality of the farming environment. However, quality assurance is no more than advertising unless it is backed up by effective, independent quality control. Chapter 9 describes the construction, implementation and promotion of quality control procedures for animal farms, necessary to provide sound evidence in support of QA claims, to identify areas for improvements in husbandry and to justify higher costs for products of higher value. The immediate aim is to create incentives towards improved husbandry by rewarding farmers for doing things better. The longer term aim is to encourage us consumers to give greater respect to food of animal origin, to spend the same proportion of our income on the purchase of less meat, milk and eggs, thereby improving our own health and helping to encourage fair shares for all.

Chapter 7 explores the 'high science' approach to improving efficiency and reducing the environmental impact, the 'long shadows' of future animal production. This includes the application of genetic and molecular techniques to the manipulation of the food animals, especially in the context of resistance to disease; and the manipulation of feeds for the food animals to increase the availability of 'complementary' feeds, reduce pollution and the threat of global warming. The intention is not to describe new laboratory techniques such as genomics in any detail: rather to review, without prejudice and from first principles of science and ethics, how these new procedures may rate on grounds of feasibility, applicability, consumer appeal and respect for the animals.

Chapter 8 develops the theme of 'Planet husbandry'. Here the business of animal production for food and fibre is considered as just one contributor to land management within a broad context that includes the generation of energy from biomass, biofuels or wind farms, the use of pasture and forestry as carbon sinks, and a profitable role for grazing animals in the management of amenity and conservation areas.

The final chapter will explore how ideas for a better future for animal farming can be put into effect through the mobilisation of public opinion, on the basis that the responsibility of care for managing the animals and the land lies not just with the farmers and the landowners but with us all. These options for husbandry regained have been based on a healthy respect for first principles but they are absolutely not a didactic exposition of what I believe to be right. Nevertheless it is proper to consider best ways to increase public demand for 'better' products and practices, where 'better' is defined not only by conventional criteria such as appearance, taste and price but also by less immediate measures of value such as animal welfare, fair trade, sustained quality of the land and quality of life for those that live on the land. The paths to change may involve legislation, by prohibition, prescription or incentive, the stick and the carrot. However, legislation is slow and legislation by prohibition or prescription can never do more than establish baseline standards, even though, from time to time, the bar may be raised (viz. the 2012 imposition within Europe of improved minimum standards for cages for laying hens). Legislation by incentive is inherently more attractive because it is more flexible. Yet when it involves expenditure by governments the incentives are likely to be small (viz. the cross-compliance payments for animal welfare and environmentally sympathetic farming from the European Community).

While it is true that legislative changes in democratic societies arise in response to public opinion, there is good evidence that a more effective route to change is through the expression of the individual opinions of those who believe they have a good cause, and the spread of these opinions among an increasing circle of believers (viz. the explosive increase in demand for free-range eggs in the UK, from under 5 per cent to over 50 per cent within 15 years). This 'politics by other means' is undoubtedly effective but it can be dangerous. I have already cautioned against the practice of arguing from limited premises and this is seldom more apparent than in relation to issues of animal welfare and the environment where, in the words of Yeates, 'the best lack all conviction while the worst are full of a passionate intensity'. Emotion is an essential weapon in the pursuit of the good, but it is one that should be handled with care and only after due thought. In line with Buddhist doctrine, emotion is best directed when cerebral, not visceral.

Note

1 *Chambers Dictionary* gives one definition of the milch cow as 'a reliable source of income'.

2 Audits of animals in agriculture

The poison is in the dose.

(Paracelsus)

This chapter is mostly about numbers. It begins simply but gets more complicated later. It is self-evident that the production of food for human consumption from animals is, in overall terms, less efficient than the production of food direct from plants because only a proportion of the feed they eat is converted to the food we eat. (Conventionally we say that humans eat *food*, farm animals eat *feed*.) The first section of this chapter will examine the factors that determine the efficiency of conversion of animal feed to human food. It is also a matter of increasing concern that livestock farming is a major contributor to environmental pollution, including the minerals nitrogen (N) and phosphorus (P) and the greenhouse gas methane. These concerns will be addressed using the principles of 'Life Cycle Assessment' which is 'a standardised accounting framework used to inventory the material and energy inputs and emissions associated with each stage of a production life cycle' (Guinee *et al.* 2002).

Before getting into detail, however, it is useful to remind ourselves of the central contention of Chapter 1: farming animals is not necessarily a bad thing. In many areas of the world, from the steppes of central Asia to the Arctic, animals are the only option. Elsewhere, most traditional farming systems were mixed systems, where the animals made an essential contribution to the output and sustainability of the overall enterprise. The proper relationship between plants and animals is essentially symbiotic. Most animals (especially farm animals) depend on plant materials for nutrition. At the same time, plants need animal emissions: carbon dioxide (CO_2) for photosynthesis, N, P and K (potassium) for growth. In a sustainable system the C, N and P cycles are in equilibrium. Problems arise (and they have) when these equilibria are upset. Most sources of agricultural pollution can simply be attributed to too much potential plant feed in the wrong place: 'the poison is in the dose'.

Nutrient needs

We animals depend for our life on energy and materials that we obtain from plants and other animals; energy to fuel the fire of life, materials to build and maintain our bodies. In quantitative terms our need for energy dominates our nutrient requirements; almost always greater than 80 per cent, and for adults neither growing nor lactating, about 90 per cent. Energy is captured in the first instance by plants and converted into organic matter through photosynthesis. The main sources of chemical energy for animals are the carbohydrates, made up of carbon (C), hydrogen (H) and oxygen (O), fats and oils (predominantly C and H). The next nutritional need of animals is for protein (C, H, O and N); or strictly speaking, amino acids derived from the digestion of plant and animal proteins then used as building blocks for synthesis of body tissues, such as muscle and for export proteins such as milk and eggs. Some amino acids (e.g. lysine, leucine, tryptophan, methionine) are defined as *essential* because they cannot be synthesised in the body and must be provided in the diet. Adult humans can thrive on a diet containing no more than 12 per cent protein, provided that this protein is balanced with respect to our needs for essential amino acids. Even the most productive animals measured in terms of their capacity to export proteins as food, the dairy cow and the laying hen, are seldom fed a diet containing more than 20 per cent protein. It should be obvious that proteins from other animals are likely to have a higher biological value than plant proteins defined in terms of the essential amino acid needs of the human animal. The biological value of plant proteins is very variable. The importance of soya beans in the diet of the food animals (and the diet of those humans who don't eat animals) lies in the fact that they are almost as well balanced with respect to essential amino acids as an 'ideal' protein such as milk casein. Animals have other specific and essential material dietary needs, namely the vitamins and minerals (e.g. sodium, potassium, calcium, phosphorus, sulphur (Na, K, Ca, P and S). These in total only contribute about 2–4 per cent of nutrient requirement.

The destiny of dietary energy and materials consumed by animals is illustrated in Figure 2.1. Animals consume a diet derived from various sources. In Figure 2.1 the feeds that make up the diet are cereals, forage (e.g. hay or silage) and soya bean meal, the last fed primarily as a source of extra protein. The diet will also probably be supplemented with a small quantity of vitamins and specified minerals so as to match nutrient supply to nutrient requirement.

This mixture of feeds is then subjected to digestion. In simple stomached animals like humans and pigs, relatively simple carbohydrates (sugars and starches), proteins and fats are broken down by the processes of digestion into smaller compounds (e.g. glucose, amino acids, fatty acids) which are absorbed from the gut and become 'available nutrients' (available for metabolism). Material that cannot be digested (C, N, P, etc.) is eliminated in the faeces. (At this stage I must apologise to proper nutritionists. This is all very superficial, but necessary to what follows.)

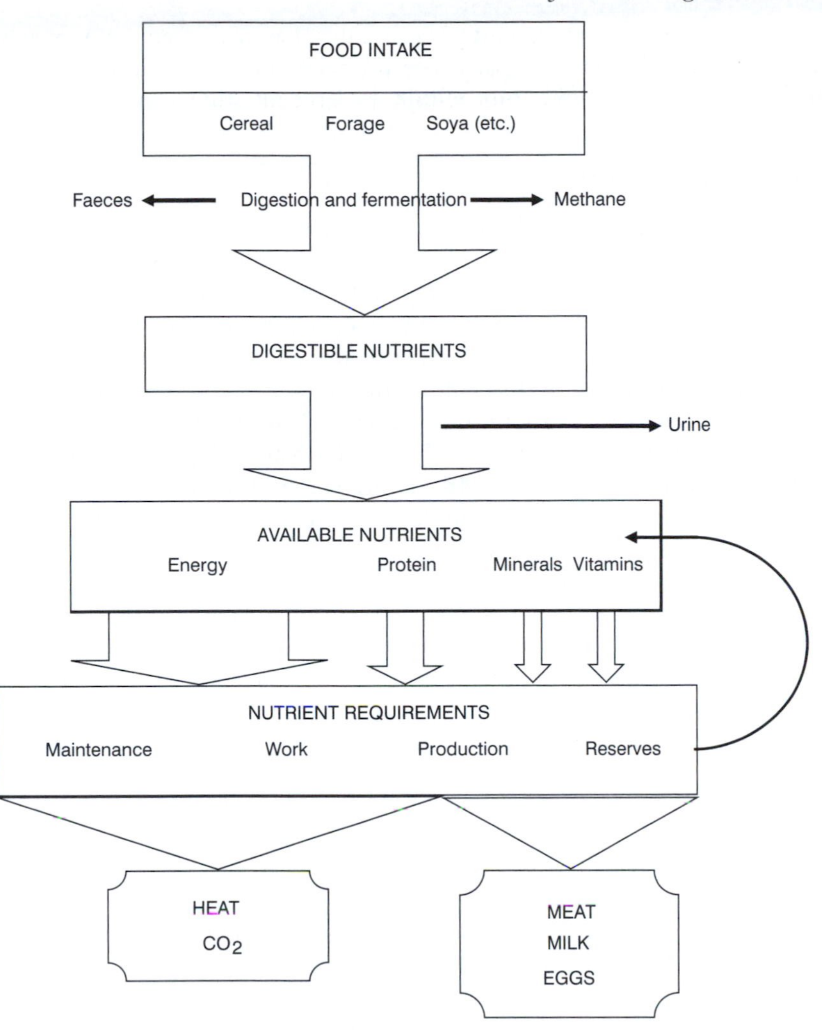

Figure 2.1 Nutrient supply and requirement

Complex structural elements of plant fibre (e.g. cellulose) cannot be broken down by digestion in the stomach and duodenum of simple stomached animals. They can however be fermented by microorganisms in the anaerobic environment of the rumen and large intestine to salts of short-chain, volatile fatty acids, principally acetate and propionate. This process leads to the emission of methane in amounts determined primarily by the acetate:propionate ratio. The extent to which plant fibres can be fermented to yield available nutrients is primarily determined by the volume of the fermentation compartments and the length of time the material resides within them. Thus horses and rabbits can obtain much more nutrition from fibre than pigs or humans because their large intestine is much larger relative to body size. Ruminant animals like cattle and sheep can obtain even more, partly because they have a fermentation vat at each end of the digestive tract and partly because retention time within the rumen is much longer than retention time within the large intestine.

Energy metabolism

The nutrients made available by the processes of digestion and fermentation are used to meet the metabolic requirements of the animal. The first metabolic need is that of *maintenance*: the energy and materials necessary to sustain the essential processes of life in an animal neither gaining nor losing body mass. Feed consumed in amounts greater than maintenance requirement is available for conversion into animal product (meat, milk and eggs). At maintenance, nearly all nutrients are used as a source of metabolizable energy (ME) and oxidised to generate heat and CO_2 (Figure 2.1). For the farm animal, producing milk, eggs and meat nutrients other than ME acquire greater importance, especially the amino acids that will be retained in the active tissues of the growing body (e.g. muscle, skin, internal organs), or synthesised into export proteins such as milk and eggs. Thus in production rations for farm animals the concentrations of dietary protein, vitamins and minerals are increased relative to ME in proportion to the level of production (conventionally expressed relative to maintenance).

It is conventional to express the efficiency of conversion of animal feed to animal product on a weight-for-weight basis (e.g. body or carcass weight gain/ feed consumption, or feed day matter (DM) consumption). This audit of the efficiency with which feed for animals is converted into food for humans will be based on the efficiency of utilisation of ME. The reasons for this are:

- Energy exchanges in farm animals have been described with great precision (Blaxter 1989; National Research Council (USA) various dates).
- Metabolic requirements of animals for energy (and the commercial cost of providing energy as a nutrient) are very large relative to all the other nutrients.
- While feeds differ greatly in their capacity to provide ME for reasons largely linked to their digestibility, requirement for ME and the efficiency of utilization of ME are essentially functions of the animal, not the feed.

The last of these three is the most important since it allows one to compare the efficiency of different animal production systems in a way that is not confounded by differences between feeds. For a comprehensive description of factors influencing energy metabolism in farm animals see Blaxter (1989).

Maintenance of metabolically active tissues (of which the most active are intestinal epithelium, liver, muscle and nerve) requires a continuous turnover (synthesis and catabolism) of proteins and other complex compounds, and this inevitably leads to some excretion of other materials, primarily N in urine, but also other elements such as P, Na, K and S. Protein fed in excess of minimal requirement for essential processes of turnover is used as an energy source.

However, protein oxidation is incomplete and leads to the excretion of increased amounts of N as urea in mammals or uric acid in birds. Thus the more protein an animal eats relative to requirement the greater the excretion of N; a fertiliser in small doses but a pollutant when excreted to excess.

The next category of nutrient requirement listed in Figure 2.1 is *work*. In the context of animal production, work is defined as that done in excess of the requirement for maintenance, e.g. measurable work done by draught animals pulling ploughs or carts. Here again, the most important nutrient requirement is for ME; all the ME from carbohydrates and fats is emitted as heat and CO_2. Increased oxidation of proteins will lead to increased emission of N.

As indicated above, production of milk, eggs and meat requires relatively higher concentrations of protein and other nutrients relative to ME. However, the metabolic processes involved in the net synthesis of milk, eggs and meat also carry an energy cost. This is illustrated in Figure 2.2, which plots ME intake against energy retention (RE) for a lactating cow and a growing beef animal. In the case of milk production, the *net* efficiency of utilization of ME (=RE/ME) both for maintenance and for milk synthesis is 70 per cent: in other words 30 per cent of ME fed in excess of maintenance requirement is dissipated as heat. The net efficiency of meat synthesis in growing animals varies with species and diet. In pigs it is close to 70 per cent. In growing cattle and sheep it varies between 40 and 60 per cent depending largely on the ratio of simple carbohydrates to fibre in the diet: expressed simply, the ratio of cereals to forage. The *gross* efficiency of utilization of ME (RE/ME) for production of milk, meat or eggs is zero until ME intake exceeds maintenance requirement. Then it increases in curvilinear fashion at a rate depending on net efficiency and ME intake with respect to ME intake at maintenance (ME_m) to a limit determined by maximum ME intake. A high-yielding dairy cow fed a nutritionally rich diet can consume ME at a rate four times its maintenance requirement ($4 \times ME_m$). In these circumstances the gross efficiency of utilization of ME can approach 60 per cent. The cost of this exceptional metabolic demand to the high-yielding dairy cow is considered later. Growth in beef cattle is a more leisurely affair. Metabolic demand for growth is less than that for lactation. In consequence appetite is also less: maximum ME intake is less than $3 \times ME_m$ and gross efficiency slightly less than 30 per cent (Figure 2.2).

Available nutrients consumed in excess of requirement for maintenance, work, production of meat and eggs, and growth of the (essential) lean body mass are stored, mostly in the form of fat. These reserves are of course essential in times when nutrient demand exceeds supply, for example, the animal at pasture in the winter or the dry season, the high-yielding dairy cow in early lactation. Moreover the healthy farm animal has a greater need for complex proteins (etc.) than that defined by the things the farmer can sell (meat, milk and eggs). These include sufficient nutrients to sustain the integrity of the immune system and 'unsaleable' tissues such as the skin and digestive tract. Undernourished animals do not just lose fat and muscle; they also lose their resistance to infection and injury.

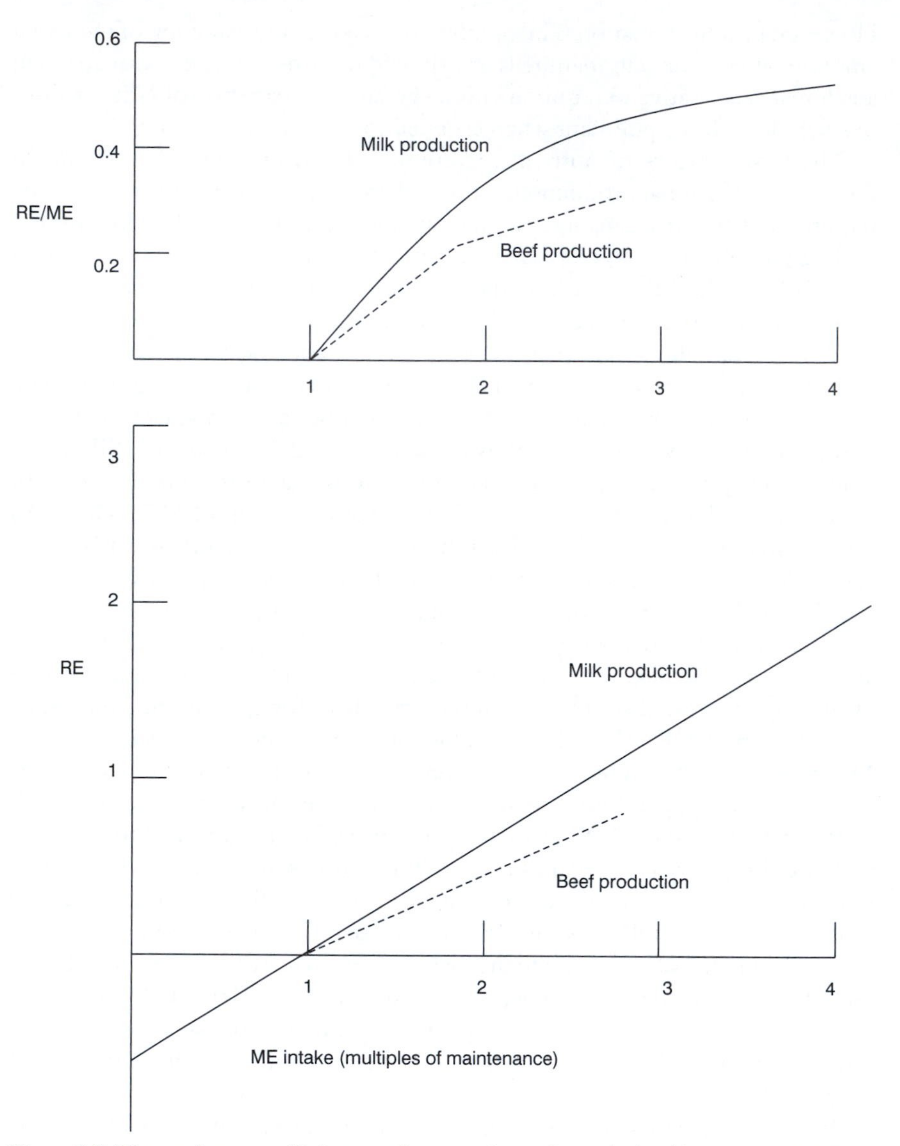

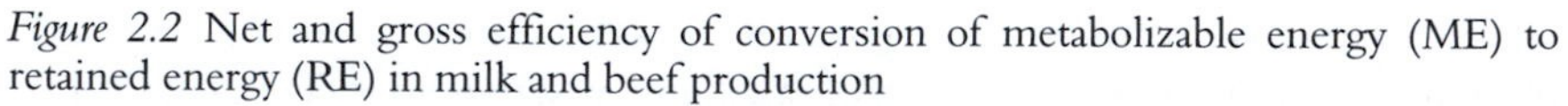

Figure 2.2 Net and gross efficiency of conversion of metabolizable energy (ME) to retained energy (RE) in milk and beef production

Energetics of maintenance and growth

Let us now examine from first principles the factors that determine the efficiency of utilization of ME for meat production. Common sense suggests that pig and poultry meat production when carried out in large intensive units are far more efficient than pastoral systems for production of beef and lamb. Moreover, these systems appear to have made far more progress in increasing both output (growth rate) and efficiency, measured as output of saleable meat relative to input of feed, than beef and sheep production. All this is undoubtedly true. It is most apparent in the case of broiler chicken and turkey production. Poultry meat, which 60 years ago was considered a luxury, is now the cheapest meat on the market, typically retailing at a price per kg less than that of canned dog meat. It is necessary however to examine in more detail the reasons why this has come about. In a growing animal, nutrients only become available for meat production after the animal has met its requirements for maintenance (Figure 2.2). Feed energy requirements for maintenance can be minimised by housing growing pigs and chickens (the slaughter generation) at a high stocking density in environmentally controlled buildings, so as to minimise energy requirement for activity and eliminate energy requirement for the maintenance of homeothermy. In these circumstances the gross efficiency of conversion of feed to saleable product can exceed 50 per cent in broiler chickens and 40 per cent in growing pigs, whether measured as RE/ME, or more practically as kg carcass produced per kg feed. However any audit of the efficiency of feed conversion in an animal production system needs also to consider the feed required to support the breeding generation, whose direct contribution to food supply is likely to be very small.

Table 2.1 examines the relative contribution of the breeding and slaughter generations to energy demand in four meat species: broiler chickens, pigs, sheep and beef cattle. A broiler breeder hen weighing 3kg produces 240 viable offspring per annum that can be fed to yield 1.5 kg of saleable meat in about 40 days from hatching. In these circumstances one broiler breeder can produce 120 times her own weight in carcass yield per annum. The proportion of total feed energy consumed by the breeding generation is only 4 per cent of total input, 96 per cent is fed to the slaughter generation and converted at high efficiency into poultry meat. Thus when the feed requirements of both the slaughter and breeding generations are taken into account, overall efficiency of conversion

Table 2.1 Allocation of feed energy to the breeding and slaughter generations in broiler chickens, pigs, sheep and suckler beef cattle (Webster 1989)

	Broilers	*Pigs*	*Sheep*	*Beef cattle*
Weight of breeding females (kg)	3.0	180.0	75.0	450.0
Carcass weight from meat animal (kg)	1.5	50.0	18.0	250.0
Ratio, weight carcass/weight dam	0.5	0.3	0.2	0.6
Progeny/year	240.0	22.0	1.6	0.9
Weight carcass per year/weight dam	120.0	6.2	0.4	0.5
Proportion of food energy/year				
To breeding generation	0.04	0.20	0.68	0.52
To slaughter generation	0.96	0.80	0.32	0.48

of animal feed to saleable carcass can still exceed 50 per cent (Webster 1989). The breeding sow in this example produces 22 piglets per year and the ratio of carcass yield to sow weight is 0.28. In this case 20 per cent of feed energy is used to support the breeding generation, 80 per cent goes to the slaughter generation. In the case of sheep and suckler beef cattle, the numbers of progeny per year are 1.6 and 0.9 respectively. Carcass yields are 0.38 and 0.5 maternal weight, with the consequence that only 0.32 and 0.48 of feed is used to grow the lambs and young cattle sold for prime meat. Clearly the main factor influencing the overall efficiency of the enterprise, when measured strictly in terms of energy conversion, or feed conversion, is the prolificacy of the breeding females.

Growth

The next thing to take into account when evaluating from first principles the relative efficiency of different systems of meat production are some fundamental rules regarding the efficiency of growth (Webster 1989). Figure 2.3 illustrates the basic principles that govern the efficiency of conversion of feed energy (ME) to weight gain in any continuously growing mammal. As the animal grows from weaning (W) to maturity (M), ME intake increases and so too does heat production (H), until at maturity, when the animal is at maintenance ME=H (Figure 2.3a). Energy retention (RE=(ME-H)) peaks at 20–25 per cent M, then declines rapidly (Figure 2.3b). Also, during the process of continuous growth to maturity there is a progressive increase in the proportion of energy retained as fat (at 39MJ/kg) relative to true lean (protein and water in muscle at 4.5MJ/kg), with the consequence that weight gain per MJ declines at an accelerating rate (Figure 2.3c). Feed conversion efficiency (g gain/MJ ME, Figure 2.3d, derived from Figures 2.3b and 2.3c) is relatively constant up to the stage of peak energetic efficiency (*c.*0.25M) then also declines at an accelerating rate. In the example illustrated by Figure 2.3 it is most efficient to slaughter the animal at about 25 per cent of mature size.

These basic laws of growth do much to explain why genetic selection for growth in broiler chickens and pig production has been able to achieve such spectacular increases in growth rate and feed conversion efficiency. Most of the increase in both growth rate and feed conversion efficiency can be simply attributed to selecting for animals at a greater mature weight and killing them at a lower stage of maturity. This should be apparent from inspection of Figure 2.3. If not, consider the *reductio ad absurdum* of this argument. The way to achieve maximum feed conversion efficiency in the slaughter generation would be to kill the animals at birth before they had time to eat anything. The snag with this approach is, of course, that there would be very little to sell, certainly not enough to justify the cost of feeding the breeding generation all year.

In pigs there has been a significant trend to produce animals with more lean and less fat, partly in response to consumer demand and partly to increase feed conversion efficiency, since pure lean (fat-free muscle) contains only about 12 per cent the amount of energy in pure fat (4.5/38). Some of the increase in

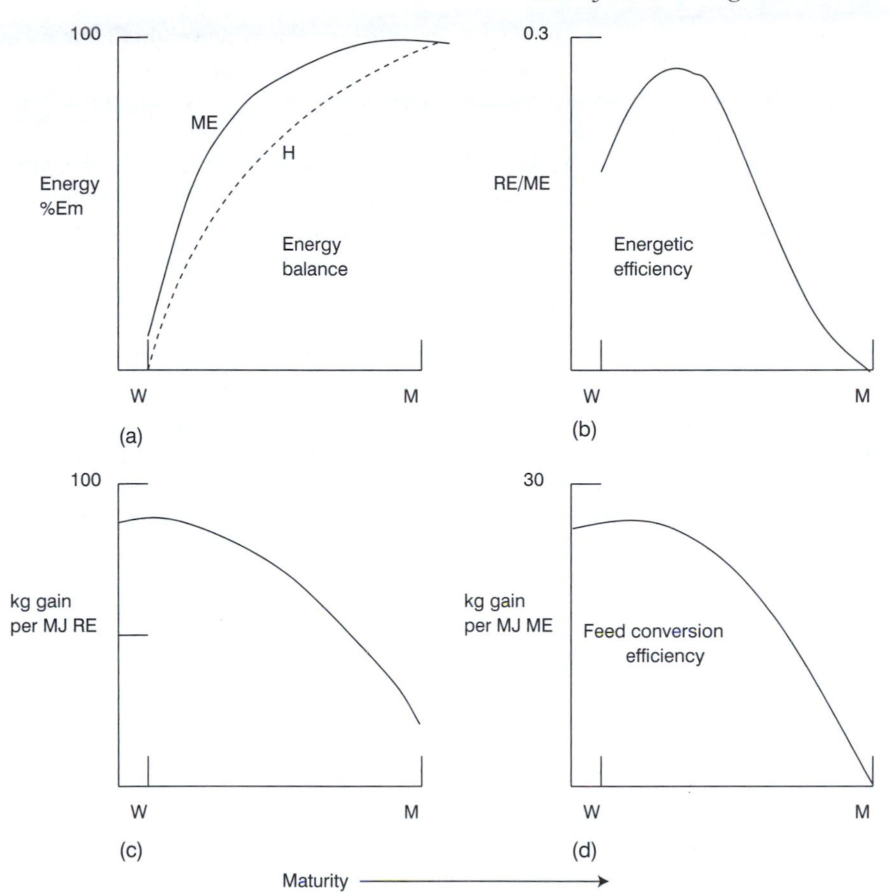

Figure 2.3 Factors affecting the efficiency of utilization of metabolizable energy (ME) for growth in mammals. RE is energy retention, H is heat production. W and M are weights at weaning and maturity. E_m is maintenance energy exchange (ME=H) at maturity. (Webster 1989)

leanness in pigs can be attributed to genetic selection for animals that are leaner at any age. However, much can be attributed to the policy of killing the animals at a lower proportion of mature weight (Figure 2.3c). One important effect of this breeding policy has been that mature breeding sows and broiler hens are getting bigger and bigger, with some deleterious consequences for their own welfare (see Chapter 4). In strictly commercial terms, selection for bigger and bigger parents presents few problems for producers because the breeding sows only consume 20 per cent of total feed, and broiler breeders only 4 per cent.

Consider now what would be the consequence of selecting exclusively for bigger and bigger sheep. Table 2.1 shows that 68 per cent of feed is eaten by the breeding generation. Moreover, the breeding ewes and rams have to be fed all year. Lambs are typically born in the early spring, in anticipation of the flush of nutritious, cheap grass and sold, ideally after about three to four months before the grass runs out and at a relatively low proportion of mature weight (Table 2.1). A policy to select all sheep simply on the basis of mature size would *decrease* the overall efficiency of the enterprise. Sheep farmers in the United Kingdom, who are more canny than many scientists perceive, address this problem through a policy of *divergent selection* (Webster 1989). The slaughter generation of lambs are typically a three-way cross. The mothers of the majority of prime slaughter lambs are typically a first-generation cross between a small, hardy hill sheep (e.g. Scottish Blackface, with a low maintenance requirement) and an 'improver' ram (e.g. Blueface Leicester) to increase prolificacy. This hardy, prolific ewe, exhibiting hybrid vigour, is then crossed with a large, fast-growing ram (e.g. Texel or Suffolk) to maximise growth rate and slaughter weight in the slaughter generation. In simpler words, the optimal breeding strategy for sheep is based on small, hardy, prolific mothers and large, lean, muscular, fast-growing fathers. In the context of the present argument, the effect has been to minimise the proportion of feed necessary to support the breeding generation relative to that fed to the slaughter generation. In a much broader context, the 'stratification' of the British sheep industry, whereby small, hardy mothers are bred on marginal hill and upland pastures, and most of the lambs are grown on the best lowland grass, represents one of the most elegant examples of biologically efficient and ecologically sensitive animal production in the world – though, like too many ecologically sound systems, it doesn't make enough money.

A similar policy of divergent selection applies to beef production by mating large bulls (e.g. Charolais) to small suckler cows (as with sheep, usually cross-bred animals displaying hybrid vigour for maternal traits). Their role is to produce one calf per year (increased prolificacy is not an option). Beef production from the dairy herd is completely different since the lactating (Holstein) cow is the major earner. Table 2.1 shows prime beef animals being slaughtered at a carcass weight 55 per cent that of their mothers. This is partly achieved by divergent selection (i.e. large fathers). It is also achieved by growing the animals relatively slowly to reach slaughter condition at two years or more. This is because the proportion of fat in body gains increases with increasing growth rate (RE/ME). A meat animal is said to be 'finished' when it reaches the degree of fatness required by the butcher. Thus lambs growing and fattening at a high rate over one grazing season are slaughtered at a much lower proportion of mature weight than cattle grown slowly over two or more seasons (Table 2.1).

Energetic efficiencies of milk, meat and egg production

Milk and egg production appeal to vegetarians and to thrifty souls like myself because they do not involve killing the animal to get the food out. Milk and eggs are both total foods of the highest nutritional value: the former to feed the growing calf from the time of birth, the latter to feed the fertilised embryo up to the time of hatching. Table 2.2 compares the efficiency of conversion of feed energy (ME) and protein into hens' eggs, cows' milk, pork meat from intensively reared pigs and beef from extensively reared cow-calf systems where

Table 2.2 Efficiency of energy and protein conversion in meat, milk and egg production, described by the ratio of output to input. Output is defined by energy and protein in food for humans; inputs in terms of total and 'competitive' intake of ME and protein, where 'competitive' describes energy and protein from feed sources that could be fed directly to humans (Webster 1994)

	Eggs	*Pork*	*Milk*	*Beef*
Production unit	1 hen	22 pigs	1 cow	1 calf
Support unit	0.05 hens	1 sow	0.33 heifers	1 cow
Output/year (kg food)	15	1300	8000	200
MJ food energy	130	13000	28000	2500
kg protein	1.65	208	264	32
Input/year (MJ ME in total)	389	67038	67089	29850
MJ 'competitive' ME	351	53630	20127	10268
Input/year (kg protein in total)	5.2	818	946	361
kg 'competitive' protein	5.0	736	236	108
Efficiency				
Food energy/total feed ME	0.33	0.19	0.42	0.08
Food energy/'competitive' feed ME	0.35	0.24	1.39	0.24
Food protein/total feed protein	0.32	0.25	0.28	0.09
Food protein/'competitive' feed protein	0.33	0.28	1.12	0.30

the contribution of the breeding beef cow is only one calf/year, plus her own carcass at eventual slaughter.

In each column the efficiency of conversion of feed energy and protein is expressed in two ways:

- Overall efficiency: food energy and protein (for human consumption) relative to total feed energy and protein consumed by the animals (both the breeding and slaughter generations).
- 'Competitive efficiency': food energy and protein (for human consumption) relative to animal consumption of ME and protein from 'competitive' feed sources (i.e. food that could have been directly used for human consumption) as distinct from 'complementary' feeds – e.g. forages like grass and by-products remaining after preparation of food and drink for human consumption (e.g. maize gluten, brewers' grains).

The *overall efficiencies* of energy conversion for egg, pork, milk and beef production are 0.33, 0.19, 0.42 and 0.08 respectively; for protein conversion they are 0.32,

0.25, 0.28 and 0.09. The reason that the energy (but not protein) efficiency of milk production is greater than that for egg production can be attributed to the fact that cows can be selected to produce more and more milk, whereas chickens are still restricted to one egg per day. Both egg and milk production are more efficient than intensive production of pork meat: beef production (by these measures) fails to achieve an efficiency of 10 per cent.

Examined in terms of *competitive efficiency* of energy use the picture changes (protein will be considered later). Here beef production becomes as efficient as pork (or no less inefficient) and dairy farming becomes very efficient indeed. The values for milk production are taken from the dairy farm owned by my immediate neighbour in south-west England, classic pasture-based dairy country, where approximately 65 per cent of total ME supply is produced on farm. In these circumstances, the output of food energy for human consumption is 39 per cent greater than their demand for food that we could eat ourselves.

Table 2.2 provides a powerful illustration of the danger of leaping to simplistic conclusions based on limited or selected evidence as to whether particular animal production systems are 'better' or 'worse'. Even within these limited parameters of feed energy conversion, there is no simple answer. When other factors such as environmental sustainability, fuel costs, pollution, animal welfare and ethics are taken into account (as they will be), our view of what's best becomes decidedly fuzzy. Try to produce goods of value while doing as little harm as possible becomes a more realistic ambition.

How hard do farm animals work?

This is a neat point at which to introduce the issue of farm animal welfare into this audit of energy exchanges. One of the most serious welfare concerns is that, with the aim of increasing productivity, farmers may be working animals too hard, and breeders may be placing too much emphasis on selection for production traits at the expense of traits relating to fitness, leading to problems such as increased mortality, reduced fertility and painful problems such as lameness. These topics have recently been reviewed in detail by Jacky Turner (2010).

Table 2.3 compares the daily food energy requirements (expressed as ME), energy expenditure as work (H), 'food' energy outputs as meat, eggs or milk. The sedentary adult human (e.g. office worker) is taken as the basis for comparison (ME intake and work output as heat = 1.0). Energy exchanges of all other classes are expressed as multiples of this standard sedentary human (adapted from Webster 1992). The table compares the ME intake, energy output as 'food' and work rate (heat production, H) of some more or less active humans and farm animals. In order to avoid complications involved in scaling for different sizes ME, RE and H are all expressed relative to ME requirement for maintenance (I,ME_m where I,ME=H). In the case of the human animal, this corresponds to the sedentary adult, neither gaining nor losing weight.

Relative to most farm animals, most humans don't work very hard. Even coal miners in the 1950s, with relatively limited access to machinery, only

Table 2.3 How hard do animals work?

Species	*Activity*	*Energy exchange*		
		ME intake	*Work/heat*	*'Food'*
Human	Sedentary	1.00	1.00	
	Working miner	1.25	1.25	
	Lactating woman	1.53	1.28	0.25
	Endurance cyclist	2.60	2.90	–0.30
Pig	Grower	2.10	1.30	0.80
	Lactating sow	3.20	1.73	1.47
Birds	Broiler chicken	2.10	1.18	0.92
	Laying hen	1.73	1.30	0.43
	Passerine feeding chicks	3.03	3.03	
Cow	Suckler with one calf	2.22	1.32	0.91
	Dairy cow, 50l/day	5.68	2.14	3.53

worked about 25 per cent harder than office workers. A few humans do work spectacularly hard, for short periods. These include polar explorers and cyclists in endurance races like the Tour de France. Suckler cows, growing pigs and, surprisingly, layer hens don't work very hard (H about 30 per cent above maintenance). Hens that lay one egg almost every day for a year, do suffer from 'layer fatigue', especially when housed in battery cages, but this is not due to excessive energy demands; it is primarily a form of osteoporosis due to a failure to sustain calcium balance (Whitehead 2004).

The most energy-demanding of all metabolic processes is lactation. A highly selected Holstein dairy cow producing 50 litres milk/day may consume over five times ME_m, and work (produce heat) at a rate over twice that of an animal at maintenance. This work rate exceeds any human example but the endurance cyclist. Moreover the need to consume very large amounts of feed to meet the metabolic demands of lactation imposes a severe strain on the capacity of the cow to process feed by fermentation and digestion. The only animal species known to work harder than dairy cows are passerine birds during the period when they are catching and bringing food to their young for up to 16 hours/day. The energy demands on a breeding sow with 12 piglets are also high but, as with the passerine birds, the duration of this period of very high energy demand only lasts a few weeks. High-yielding dairy cows are expected to sustain high energy demands for ten months in every year, and exceptionally high energy demands in the first three months of every lactation. It is not surprising therefore that over 50 per cent of dairy cows fail to sustain a working life beyond three years of lactation and that most problems arise within the first three months of lactation (Pryce *et al*. 2004).

A final thought. Motherhood is hard work for all species. The energy demands of lactation on the human mother feeding one slowly growing child may be low relative to the sow, cow or swallow. However, she still works harder than a coal miner.

Efficiency of protein use

One of the most common arguments made in defence of the consumption of foods of animal origin (meat, milk and eggs) is that it provides us with essential nutrients, of which the most important is high-quality protein. This may be a valid argument in the case of a child in sub-Saharan Africa, supplementing a carbohydrate-dominated diet with one or two eggs per week. It is also a valid argument for pastoral nomads or Inuits for whom food of animal origin may be the only food available. It is seldom, if ever, a valid argument for those in the developed world. We eat foods of animal origin, high in protein and fat, primarily because we enjoy them. However, this is not just decadence. The motivation to consume energy as fats or oils (as distinct from carbohydrate) appears to be programmed into our unconscious physiology. Studies on food choice in affluent areas of Europe and the USA suggest that we regulate our diet (largely unconsciously) to provide 35–40 per cent of ME in the form of fats and oils (Bludell *et al.* 1996). For omnivores and ovo-lacto vegetarians much of this comes from meat, milk or eggs. Even vegans appear highly motivated to select foods with a relatively high ratio of vegetable oils to carbohydrates (e.g. biscuits rather than bread). Racial groups such as the Chinese and Japanese have in the past been commended by nutritionists for selecting low-fat diets relative to those in the decadent West. However, with increasing affluence, it appears that they are becoming just the same as us when assessed according to the proportion of fats/oils in the diet (Bludell *et al.* 1996). I shall return to this issue in Chapter 4.

One of the most common accusations made in the assault on the consumption of foods of animal origin is that vast areas of land (e.g. the denuded Brazilian rain forest) and vast resources of water and fossil fuels have been directed to the production of protein-rich feeds for livestock. The most severe criticism has been directed at feedlot beef production, largely on the basis of the volume of land involved and (as illustrated above) the relative inefficiency of beef production. In the case of soya production, this accusation is broadly true, although the soya bean meal fed to livestock is the residue after removal of soya oil, largely for human use. (Soya bean meal contains only 30 per cent of the energy in the original bean.) As indicated earlier, soya bean meal is the preferred protein supplement for simple-stomached animals like pigs, poultry or human vegans because the balance of essential amino acids is very close to animal requirements.

Soya bean meal is a nutritionally attractive, and currently cost-effective protein feed for ruminants. However, it is entirely unnecessary. To explain this it is necessary to give a brief description of protein digestion in the ruminant.

Most of the protein consumed by ruminants, whatever the source, is completely degraded in the rumen to ammonia, which is then used for microbial protein synthesis. Microbes passing out of the rumen undergo acid digestion to amino acids and small peptides, which are then absorbed across the intestinal wall and used for protein synthesis in the host animal. This is described as rumen degradable nitrogen (RDN). The rumen microbes can synthesise protein from degradable protein in the diet and from simple non-amino N compounds, principally urea. In natural circumstances the contribution of urea to microbial protein synthesis comes from blood urea recycled to the rumen in saliva. In some practical circumstances (e.g. feeds for finishing beef cattle) it can be economic to include urea or other sources of non-protein N in the diet. Thus the amino acid composition of RDN is (almost) irrelevant. Moreover the biological value (the amino acid composition) of microbial protein is comparable to that of soya. For a fuller description see Webster (1993).

The incorporation of urea into microbial protein is much more important in an ecological sense. One of the reasons that ruminants are particularly well adapted to long periods of undernutrition, e.g. during the dry season in the tropics, or the winter season at higher latitudes, is that the urea that undernourished ruminants produce as a result of metabolising their own protein reserves as an energy source, can be recycled to the rumen, converted by the rumen microbes into protein, and recycled to the host animal as essential amino acids. Simple stomached animals (e.g. humans, pigs and chickens) cannot do this to any significant extent, which is another reason why ruminants, properly managed and in the right habitat, are anything but wasteful of resources.

In recent years there has been a fashion for feeding relatively large amounts of Rumen Undegradable Nitrogen (RUN), otherwise known as bypass protein. This refers to protein that escapes degradation in the rumen and is subject to acid digestion, as in pigs and poultry. The nutritive value of RUN is obviously determined by its amino acid composition. The obsession with feeding soya to ruminants reflects the belief that feeding 'high-quality' RUN can produce significant economic gains. This belief has largely been eroded by new science, which indicates that when high-producing ruminants are fed well-balanced diets the synthesis of microbial protein can be considerably greater, and the requirement for RUN correspondingly less than that which appears in many conventional tables of feed requirements (Alderman and Cottrill 1993).

This rather complex argument is necessary to support the contention that protein feeding of ruminants need not compete with the protein needs of humans. An excellent example of this involves the feeding corn (maize) gluten as a protein source for dairy cows. Gluten is a by-product of the processing of corn for starch for a variety of human uses. The biological value of corn protein is totally unsatisfactory for human consumption because it is extremely unbalanced with respect to essential amino acids (and entirely deficient in zein) but it is a very cost-effective component of slow-release RDN in feeds for dairy cattle.

In the future, the utilization of 'complementary' protein by ruminants is almost certain to increase. The production of biofuels from oilseeds raises a stack of environmental issues that are well outside the scope of this book. Nevertheless it will happen and you cannot fuel internal combustion engines on protein. The greater the production of biofuels, the greater the generation of by-products, consisting primarily of low biological value protein and plant fibres. The tastiest solution to scavenging this waste will be to convert it into meat and milk.

Life-cycle assessment of animal systems

Life-cycle assessment (LCA) is 'a standardised accounting framework used to inventory the material and energy inputs and emissions associated with each stage of a production life cycle' (Guinee *et al.* 2002). This topic is massive in scope, critically important and consequently fertile ground for scientific research both theoretical and experimental. In this brief examination of life cycles in the context of animal farming, I have drawn heavily on a recent series of publications by Pelletier *et al.* (Pelletier 2008; Pelletier *et al.* 2010a, b) that analyse in detail the environmental costs of the major inputs and emissions involved in animal production systems. The most important inputs are energy, protein and water. It is however necessary to distinguish between (relatively) sustainable and unsustainable sources of these three resources. Energy inputs may be classified as total feed energy, 'competitive' feed energy and fossil fuel energy. The sources of feed energy itself may be subdivided into 'Free Energy' (solar energy converted into plant matter by photosynthesis) and 'Fuel Energy' used in the production, fertilisation, harvesting and processing of plant material for animal feed (Figure 2.4). Similarly, the cost of providing protein for livestock can be subdivided into protein (e.g. soy bean meal) grown mainly, though not exclusively, to feed livestock, protein available as a by-product of food grown primarily for human use (where this includes both food and biofuels), and animal protein that ruminants can synthesise from non-protein sources such as urea. Water demand is not a problem when the water is free; i.e. sustainable 'green water' that falls in abundance on green pastures: it becomes a problem when water used for irrigation of feed crops or dispersal of pollutants competes with other needs when resources are scarce.

Inputs (GJ) are assessed in terms of total feed ME, 'competitive' ME and fuel energy. Emissions of GHG (greenhouse gases) are assessed in terms of CO_2 equivalents. Most of the data have been derived from Pelletier *et al.* (2010a, b).

Total energy use

Table 2.4 presents some examples of the application of LCA to examine total energy use in livestock production systems. It has been adapted (and greatly simplified) from Pelletier *et al.* (2010a, b). The examples include three intensive ('commercial') systems, broiler chicken, pork and feedlot beef; and two more 'traditional' alternatives, 'niche' pork (equivalent to organic or free range) and pasture-finished beef production. The output of the system in every case is one

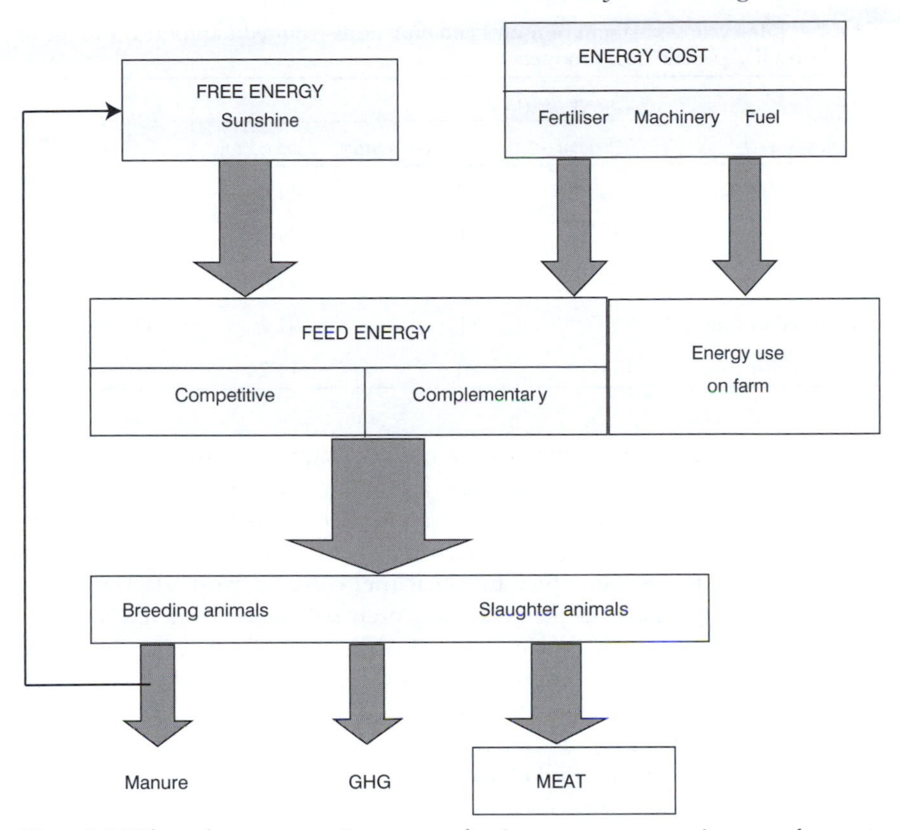

Figure 2.4 Life-cycle assessment in meat production systems: energy inputs and outputs up to the farm gate

tonne of meat. The data used to generate this table have all been drawn from units operating in the central belt of the USA. This is important because much of the demand for fuel energy derives from the production of feed crops (corn and soya) from fertilised, irrigated pastures. Clearly the demand for fuel energy in particular will vary enormously according to the extent to which animals can harvest feed for themselves, relative to that which depends on machines.

In these examples, broiler poultry production is the most efficient in terms of meat production relative to energy intake, whether measured in terms of total or competitive ME. However, it is slightly less efficient in terms of fuel energy use, attributable mainly to increased energy use in feed production and environmental control. Intensive beef production (feedlot finished) is extremely inefficient in relative terms, measured in terms of total ME and fuel energy use.

In these examples the two 'traditional' or 'alternative' systems, niche pork production and pasture-finished beef, were significantly less efficient than their intensive comparison, when assessed both in terms of feed energy and fuel energy. It is clear that one cannot make a case for alternatives to intensive livestock farming on the basis of energy cost, measured in terms of fossil fuel use.

Table 2.4 Life-cycle assessment of inputs and emissions required to produce 1 tonne of meat in broiler, pig and beef production systems

Output (1 tonne meat)	*Energy use (GJ)*			*GHG*
	ME total	*ME competitive*	*Fuel energy*	CO_2 *equiv*
Broiler chickens	36	32	14.9	1.39
Pork, commercial	52	41	9.7	2.47
Pork, niche	68	49	11.4	2.52
Beef, feedlot finished	149	51	84.2	32.7
Beef, pasture finished	194	20	114.2	45.3

Here it is important to introduce a note of caution. I am confident that the calculations presented in Table 2.4 accurately represent energy costs in the systems they describe. However, the main fuel cost is that involved in the growth of feed crops and the manufacture, distribution and presentation of animal feeds. It is inevitable therefore that fuel costs (in particular) will vary greatly according to how the feed is grown, prepared and presented. I have not attempted a comprehensive review of international publications on life-cycle assessments in animal production systems. Every evaluation is likely to be case specific so that average values derived from a comprehensive meta-analysis would not be particularly useful. My primary intention here is merely to highlight the questions we need to ask.

Water

The water footprint of a food product is measured in terms of the volume of water consumed and the volume of water polluted. Hoekstra (2010) characterises water footprint as *green* (that met through rainfall), *blue* (that produced by irrigation) and *grey* (that required to dilute pollutants to agreed water quality standards). Production of food of animal origin almost invariably requires more water per kg of food (or kJ of food energy) than the production of cereals and other vegetables. It can obviously create a problem when it competes with the water requirements of humans, but this is something that is going to vary enormously according to locality. Hoekstra presents a particularly chilling picture of the water footprint for feedlot beef production. In his example, total water required to produce 100g potatoes is 25 litres; for beef it is 1,550 litres! He acknowledges that this is an extreme example and allocates over 95 per cent of the water requirement to the production of feed (corn and soya) largely from irrigated pastures (grey water), rather than to water consumption on the beef unit itself. At the other extreme, one may consider the production of quality lamb from the green, wet uplands of the UK. In this case, the animals consume a trivial amount of the copious supply of freely available *green* rainwater and could not possibly be accused of competing with human needs. The message for water is the same as for all renewable resources: animal production only becomes a problem when water consumption, to produce feed or control pollution, exceeds that which is readily sustainable.

Methane production

One of the major environmental concerns arising from livestock production relates to the production of methane (CH_4), which is approximately 18 times as potent a greenhouse gas (GHG) as carbon dioxide (CO_2). The major 'culprits' in the matter of methane production are the ruminants (Table 2.4), or to be more specific, the microorganisms responsible for the anaerobic fermentation of plant material in the rumen. This needs a little explanation.

For practical purposes, the fermentation of all carbohydrates (starches, sugars and fibre), whether in the rumen or hind-gut, breaks these more or less complex molecules to monosaccharides (e.g. glucose $C_6H_{12}O_6$) and thence to pyruvic acid. In the simplest case this step may be written

$$C_6H_{12}O_6 \rightarrow 2CH_3COCOOH + 4H$$

The conversion of one mole of glucose to 2 moles of pyruvic acid stores a small amount of energy (2 moles ATP) for use by the microbes and generates four protons or hydrogen ions (4H). In the presence of oxygen the chemical energy contained within pyruvic acid and H are captured by the tricarboxylic acid cycle, enter the respiration chain and are oxidised to CO_2 and H_2O, with the capture of far more energy as ATP (about 36 moles per mole glucose) for use by the host animal. In the anaerobic environment of the rumen, however, the metabolism of pyruvic acid can only proceed to the salts of other short chain (volatile) fatty acids, principally acetate and propionate.

$$\text{Acetate production: } CH_3COCOOH + H_2O \rightarrow CH_3COOH + CO_2 + 2H$$

$$\text{Propionate production: } CH_3COCOOH + 4H \rightarrow CH_3CH_2COOH + H_20$$

Note that acetate production from pyruvate generates two more H ions, making six in all. Pyruvate conversion to propionate production absorbs 4H, thus net H production from the conversion of glucose to propionate is zero. Since, in normal rumen fermentation, acetate production always exceeds propionate production there is always a net generation of H. Methanobacteria in the rumen capture H and combine it with CO_2 to produce methane, which is released into the environment, mostly in the form of a belch.

$$\text{Methane production: } CO_2 + 4H_2 \rightarrow CH_4 + 2H_2O$$

Methane production is a nuisance on two counts. Not only is it a major GHG, it also represents a significant loss of chemical energy potentially available to the host animal. It is clear from the above equations that the amount of methane produced per mole of carbohydrate fermented is determined primarily by the ratio of acetate to propionate production. In theory, if it were possible to manipulate fermentation to produce only propionate, then methane production would be zero. In practice, there is always an excess of acetate production. The acetate:propionate ratio is governed largely by the ratio of fibre:starch in the diet. The greater the proportion of fibre (e.g. grass and forage crops) the greater the acetate:propionate ratio, and thus the greater the production of methane. In the grazing animal, the proportion of fermentable energy lost as methane can be over 15 per cent. On high-starch diets (e.g. beef cattle on feedlots), energy loss as methane can be reduced to about 8 per cent (i.e. halved). It follows therefore that the more 'natural' systems, such as pasture-finished beef production, generate more methane than intensive (feedlot) production, relative to output as meat, partly because of the nature of rumen fermentation and partly because slower growing ruminants ferment more feed energy before they reach slaughter weight.

This raises the Darwinian question. Why have ruminants evolved a system that appears relatively inefficient in terms of energy use? I have no satisfactory answer to this question except to say that it is probably best addressed in terms of the evolutionary fitness of the microorganisms themselves rather than that of the cow (or other ruminant) downstream. Acetate production from the relatively slow fermentation of high-fibre diets generates more energy for the microbes (as ATP) than propionate production from the more rapid fermentation of starch. Moreover fermentation of fibre to acetate is consistent with stable conditions in the rumen ($pH > 6.0$) whereas rapid fermentation of starch can lead to rumen acidosis with harmful effects on the host animal and catastrophic effects on the microbial population.

Over the last 50 years, there has been an enormous amount of research into methods designed to reduce methane production from ruminants. All these involve the selective elimination of microorganisms to increase the propionate:acetate ratio or reduce methanogenesis or both. This can be achieved, in part, by control of diet (e.g. increasing starch:fibre) or by pharmacological means. The most commercially successful of these products so far has been monensin sodium, an ion-transport inhibitor. When used in rations for feedlot cattle it significantly increases the efficiency of utilization of both energy and protein. (The explanation of the latter is outside the scope of this book; see Russel 1997.) Its efficacy tends to reduce with time and it is no longer licensed for use in the European Community for reasons relating to possible systemic effects (i.e. in the host animal, beyond the rumen). The systemic effect is such that it can kill horses. It is also pertinent at this stage to say that hind-gut fermentation in the horse emits most excess protons in the form of H rather than CH_4. Thus the direct contribution of the horse to GHG production is negligible.

I am only too aware that this, like all the forays into biochemistry in this chapter, is likely to prove confusing to those who don't know the relevant biochemistry and naïve to those who do. The above argument is presented as justification for the following conclusions:

- Ruminant animals are significant contributors to GHG production and the more extensive the system the greater the production of GHG relative to production of meat and milk.
- There is considerable capacity to manipulate the production of methane by rumen microorganisms through manipulation of diet or the use of licensed pharmaceuticals. However, dietary manipulation can create health problems for the animals resulting from unstable rumen fermentation; pharmacological manipulation can also create real health problems for some animals, and present concerns (real or apparent) to consumers and legislators anxious to ensure the production of safe, 'natural' food from animals.

Carbon sequestration in grasslands

The fashionable assault on ruminants as major contributors to global warming is perhaps the most extreme example of my especial bête noire, namely arguments from inadequate premises. For a start, it is nothing new, not even in degree. The cattle population of the USA in 2007 was approximately 100 million, of which 42 million were adult cows. In the 17th century the population of American bison was *c.*60 million. Given that feedlot animals finished on high-starch rations produce less methane than animals at pasture (per unit of ME), one can estimate, with considerable uncertainty, that methane production from ruminants in the USA is probably no more than 20 per cent greater than it was 300 years ago.

More critically, and more scientifically, it is meaningless to consider methane production in isolation. Well-managed grasslands constitute a significant carbon sink, the extent of carbon sequestration depending on factors such as the intensity of grazing and the balance between grasses and legumes (clovers and alfalfa). Recently there have been several studies of C, N and GHG balance in a wide range of grassland systems throughout Europe (Jones and Donnelly 2004; Soussana *et al.* 2010; Ramachandran Nair *et al.* 2010). Some of their most important conclusions are summarised in Table 2.5.

The first column describes a scenario where beef cattle are at pasture all year round. The pasture sequesters 471g CO_2 equivalents/m^2 per year (C units). GHG production from CH_4 and N_2O amounts to 167 C units. This system, according to Table 2.5, results in a net sequestration of 320 C units. A dairy system with cows at pasture in summer and housed in winter does result in net emission of CO_2 equivalents, although only about half that estimated from measurement of methane production alone. C sequestration from barn plus pasture exceeds that from pasture alone due to C in manure that is returned to the pasture. The final column presents figures for dairy production from cows that are housed continually. This system is more or less in equilibrium with

Table 2.5 Balance of C, N and GHG in European grassland systems. All values are expressed as g CO_2 equivalents/m^2 per year (Soussana *et al.* 2010)

	Beef, grazing	*Dairy, grazing and barn fed*	*Dairy, barn fed*
C sequestration			
pasture only	471	183	259
pasture + barn	471	269	361
CH_4 production			
pasture only	145	159	0
pasture + barn	145	387	323
N_2O production	22	64	30
Net GHG sequestration			
pasture only	+320	–22	+230
pasture + barn	+320	–163	+9

regard to GHG, largely attributable to zero production of C units from cows belching comfortably while at pasture.

Soussana *et al.* (2010) make it clear that these calculations carry a high degree of uncertainty. Moreover, they are, once again, very situation-specific and relate to well-managed pastures. Overgrazing, for example, leading to soil degradation will change the pasture from a C sink to a C emitter. Nevertheless, these calculations take a lot of heat out of the assertion that ruminants are a major contributor to global warming. Moreover, by this measure, beef cattle at pasture become the most environmentally friendly of all animals. This is not just a casual observation and I shall return to it later.

Nitrogen and phosphorus emissions from animal units

Nitrogen (N) and phosphorus (P) emitted from farm animals in urine and faeces are essential fertilisers. On the traditional mixed farm seeking to maximise food production from its own resources, whether in the developed or developing world, animal manure is a highly valuable commodity and seldom, if ever, present in sufficient quantity to optimise plant growth. It is only in the industrialised, intensive factory farm that buys in its resources and seeks to export its emissions where problems of pollution arise. As ever, the poison is in the dose.

The factors that determine emissions of potentially polluting solids, liquids and gases from intensive livestock units are illustrated in Figure 2.5. The initial production of solids, liquids and gases from animals depends (obviously) on the number of animals in the building and the amount of food they consume. Less obviously, it can be affected by the composition of the diet. Feeding high-protein rations with the aim of maximising feed conversion efficiency into animal protein increases N excretion in both urine and faeces. Increasing the proportion of fibre (non-starch polysaccharides, NSP) in the diet of pigs will reduce acid digestion,

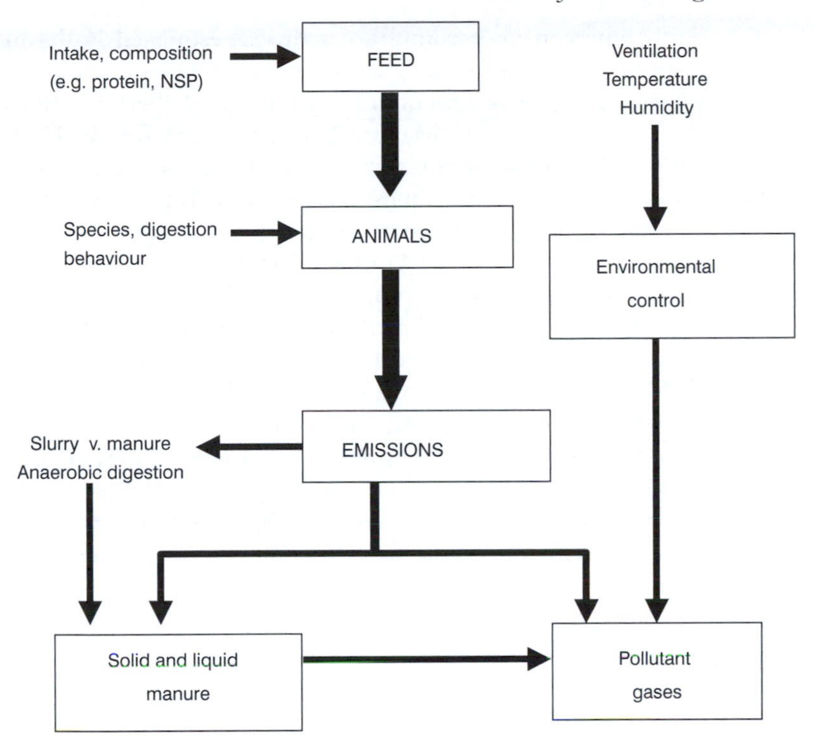

Figure 2.5 Factors affecting emissions from livestock buildings

but increase fermentation, thereby increasing the production of both faeces and methane. The fate of excreted N depends on the management of both the excreta and the aerial environment within the building. Gaseous ammonia produced by microbial action of urea, uric acid and other end products of protein digestion and metabolism is polluting, seriously unpleasant to animals and man, and a waste of potentially good fertiliser. Liquid slurry is relatively easy to spread on pastures as fertiliser. Manure, containing a high proportion of solids can be spread as fertiliser. Moreover, it is an excellent substrate for anaerobic fermentation. Installation of an industrial-scale anaerobic fermenter on an industrial pig or poultry unit can convert a high proportion of chemical waste into methane, captured and used for fuel. This is discussed further in Chapter 8.

The most extreme problems of N and P pollution from intensive livestock units have arisen in south and east Holland. The increase in pork and broiler production was made possible by a growing international demand for animal products and an EU agricultural policy that favoured the import of feeds. Between the early 1960s and the mid-1980s, the numbers of pigs and poultry increased by 10 million (450 per cent) and 50 million (125 per cent), respectively, By 1987 this created a manure surplus calculated at 16 million metric tons or 19 per cent of all manure and 75 million kg surplus phosphate (Ramachandran Nair *et al*. 2010). This manure excess led to overapplication of manure on crops and consequently emission into ground water. It was estimated in the mid-1980s that the maximum EU standard of 50 mg nitrate per litre of ground water was exceeded on 60 per cent of agricultural land in the Netherlands (Blaxter 1989). This has led to a series of 'Manure Laws', imposed first in Holland then throughout the European Community, designed to impose quotas on the production of manure (from the buildings) relative to the land on which it can be spread. In Dutch legislation, the quota for phosphate production from all animal sources is 125 kg of phosphate (P_2O_5) per hectare of land. There are cash penalties for exceeding quota but opportunities to trade between livestock units. This policy appears to have been a success. Between 1980 and 1997 the pig population of the Netherlands increased from 10.1 to 15.2 million, but waste production (calculated as P_2O_5) fell from 230 to 186 million tonnes.

The message of this section is that industrialised livestock production involving large numbers of animals on small areas of land can create serious problems of pollution. In some parts of the world (e.g. Europe) these problems have been acknowledged and controlled (in part) by legislation. It is not my purpose to discuss legislation in detail; merely to point out that sometimes it is absolutely necessary.

Pasture conservation

I would hope that by now even the most hardened antagonist to livestock farming will concede that in areas of the world most naturally suited to permanent grassland, or mixed grass and woodland, there have to be grazing and browsing animals. They are as essential to the conservation of the land as the land is essential to them. The aim of pasture management is, obviously, to conserve, and wherever possible, improve the resource. This can (and should) be assessed not only in terms of its capacity to provide feed for the animals but also in terms of a package of environmental measures that include C sequestration, water management, amenity and aesthetics (see Chapter 8).

Effective pasture management depends on maintenance of optimal numbers and an optimal balance of species to maintain the fitness of both the animals and the pasture itself. Whether in the wild, or in managed pastures this inevitably requires an effective system for management of the grazing population through a policy of culling by humans (for meat) or other natural predators. Examples of bad pasture management range from the desertification of large areas of sub-Saharan Africa, through overgrazing by owned livestock (with goats as an especial culprit) to well-meaning but catastrophic attempts in Europe to manage wildlife reserves with grazing animals that are not harvested for food, because it would not be 'natural', but left to die of starvation in a devastated habitat.

The proper management of pastures and permanent grassland, whether in the intensive dairy production areas of Europe and New Zealand or the extensively ranged steppes of central Asia, involves more than just a sensible culling policy, it requires an understanding of the importance of mixed grazing to parasite control and the different grazing behaviour and digestive physiology of the grazing animals. The aim of parasite control, particularly in extensive systems, should be to manage the parasite burden so that it does not exceed the capacity of the grazing animal to cope and maintain fitness through development of effective immunity. This can be achieved not only by control of stocking density, but also by simultaneous or successive mixed-species grazing (e.g. cattle and sheep together, or cattle followed by sheep). Pasture management for parasite control has been well described elsewhere (Krecek and Waller 2006) and will not be discussed further here.

I shall dwell at greater length on the impact of grazing behaviour and digestive physiology because it is highly important but often ignored. Cattle, and other bovid species like the buffalo and the North American bison harvest grass by scything it with their tongues over the blade of their lower incisors. When the grass is of sufficient length, this can harvest large quantities of feed in a short time, which is consistent, in survival terms, with allowing the animals to ruminate the material in a place of relevant safety, and consistent, in a commercial environment, with very high levels of production (e.g. milk production in Holstein cows). The process of rumen fermentation is also extremely efficient at extracting the maximum amount of nutrients (e.g. ME) by fermentation of plant fibre. Fermentation of plant fibre is a slow process, so the extent of fermentation is determined by the size of the fermentation compartment(s) and the length of time available for fermentation (retention time). In this regard ruminants are able to extract more nutrients per unit of plant fibre than a hind-gut fermenter like a horse. However there are two main disadvantages of being a cow. When the grass becomes very short (<2 cm) their eating behaviour (scything with the tongue) becomes increasingly ineffective. Secondly cows (especially high-performance dairy cows on rich, wet pastures) deposit their faeces in large wet cowpats that encourage grass growth but in very uneven fashion and in the form of rank grass that the cows won't eat. At the end of the grazing season an intensively managed cow pasture can look a real mess. On the other hand, the inability of bovids to overgraze pastures undoubtedly contributed to the successful colonisation of the Prairies by the North American bison.

Sheep and goats have a similar dental arrangement to cattle (incisor teeth on the lower jaw only) but are able to harvest grass down to ground level. This can be a benefit to the animals but can contribute to overgrazing if the animals are not managed properly. Goats are more successful than sheep in terms of personal survival in a nutritionally challenging environment, partly because they are more agile (they can climb trees to browse), but mainly because they are much more imaginative in terms of feed selection: they are prepared to sample almost anything. One only needs to look at sheep and goats at the end of the dry season in sub-Saharan Africa to confirm this assertion. The down side, of course, is that goats are much more destructive.

Sheep have made a significant contribution to pasture management in UK, both in pasture-based dairy farming systems and as caretakers working to preserve the aesthetics and amenity of parklands. This is partly because they crop the pastures evenly and close to the ground and partly because they wag their tails as they defaecate and distribute their small boluses of dry faeces as well as any industrial spreader of fertiliser. Thus it is common practice to move sheep on to land grazed throughout the summer by dairy cows in order to 'clean up' the pastures. It is also common practice to use sheep to preserve standards around our stately homes.

Horses (and other equids like donkeys and zebra) are less efficient than ruminants measured in terms of the digestibility of plant fibre because retention time in the hind-gut is less than in the rumen. This is not a disadvantage to the animals so long as plant material is available in some form. Horses are designed to graze for long periods, unlike ruminants which consume meals, retire and ruminate. This is consistent with species that avoid predators by running away or hiding, respectively. Having incisor teeth on both upper and lower jaws, horses are able to harvest feed to ground level, and even tear up roots. They are also able to strip the bark from trees and shrubs. At the time when pastures are under greatest pressure, i.e. winter in the northern latitudes and the dry season in the tropics, the quality of the standing grasses is very poor, containing a very high proportion of lignified (woody) grass fibre. In these circumstances the equine digestion has a higher survival value than ruminant digestion. Horses can eat very large quantities of very fibrous material, pass it rapidly through the gut, digest a very small proportion of ingested feed, but still maintain an ME intake close to maintenance. Feed intake in ruminants, on the other hand, is constrained by the rate at which they can comminute, partially digest and pass fibrous material out through the bottle-neck of the small reticulo-omasal orifice. Thus on very poor pastures a ruminant can starve to death with a rumen full of undigested fibre. Examples of this can be obtained from the inspection of body condition in cattle and ponies wintered on common land in the UK (e.g. the New Forest), mortality rates of cattle and ponies in the Oostvarderplassen wildlife reserve in Holland, and the common assertion that you never see a thin zebra!

Silvo-pastoral systems

There is increasing recognition that silvo-pastoral systems, made up of pasture and (mostly) deciduous woodland can make a major contribution to environmental management through C sequestration, management of water resources and flood control, aesthetics and amenity. Total C capture is approximately twice as great in pastures with scattered trees than in treeless pastures (Krecek and Waller 2006). It is possible to combine trees with arable farming, e.g. maize growing, but this is outwith my remit. If the system involves grazing land and grazing animals, then it needs animals that can manage the pasture without escaping or damaging the trees. On this basis cattle and sheep are more suitable than goats, horses or deer. Deer are particularly prone to damage young trees and can jump high fences. However, affluent owners of large areas of land may prefer to stock the land with deer because they prefer the look of them, they like to hunt them, or they can charge other people to hunt them.

This observation leads to the final key message to be taken from these audits of animals in agriculture. Pastoral or silvo-pastoral systems such as the stratified system of sheep production in the UK or the raising of beef cattle in the cork oak forests of Portugal (Figure 1.2) can be humane, ecologically sound and, by all accounts, thoroughly good husbandry. However, they cannot be sustained from an income source that is entirely dependent on the sale of commodity products from animals (meat, wool and leather). On the other hand, the quality of habitat cannot be sustained without a planned and economic policy of culling animals to generate income from the sale of meat, augmented in some circumstances by charging more for 'free range' food and in other circumstances (e.g. deer, game birds) by charging people to kill them. Approaches to maximising the value of (non-arable) land for animals will be considered in Chapter 8.

Conclusions

Although this chapter contains some quite complex calculations and numerical arguments, the main conclusions are quite straightforward.

- Animal production is inherently less efficient than plant production in terms of overall use of energy and materials.
- Energy, the major resource for animal production, is considered in terms of free energy (from the sun) and energy cost (chemical energy used in the production of feed and on the farm).
- Feed energy is categorised as competitive and complementary. Systems that maximise use of complementary feed (e.g. some dairy systems) can produce more food energy and protein for human use than they consume as competitive feed.
- Production systems that appear more satisfactory in terms of animal welfare tend to be more costly in terms of both feed and fuel energy.
- Emissions of N and P from livestock systems only become a problem in intensive, industrialised systems where they are likely to grossly exceed the capacity of the adjacent land to reincorporate them into plant material.
- Methane production from ruminants is a significant contributor to greenhouse gas production (GHG). However, in well-managed pastoral and agroforestry systems, there can be net sequestration of GHG.
- Feedlot beef production is both exceptionally inefficient and extremely polluting by almost every measure.
- Extensive, pastoral or agroforestry-based systems for grazing animals can be productive, humane and ecologically sound. However, they cannot be sustained from an income source that depends solely on the sale of animal products as a commodity.

Note

1 The eagle-eyed will note that these columns do not quite add up. In the absence of the original data, I must assume these discrepancies result from rounding-up errors.

3 Husbandry, health and welfare

What boots it with uncessant care, to tend the homely slighted shepherd's trade.
(John Milton, *Lycidas*)

I need, at the outset of this chapter, to restate what are and are not the aims of this book. It is *not* my aim to produce an encyclopediac list of all the major health and welfare problems of all the major farmed species, since it would inevitably be very superficial and unlikely to say anything new. It *is* my aim to highlight the most important problems of animal health and welfare arising more or less from failures in husbandry, to outline logical structures for the analysis of these problems and action to address them. The first step in this process is to gather the multitude of individual problems into a manageable number of categories that may then be addressed in a generic way while illustrating each category with important practical examples to help make the point and identify priorities for action.

Good husbandry carries an absolute duty of care. This is recognised in the oath sworn by all UK veterinary graduates on admission to the Royal College of Veterinary Surgeons: 'my constant endeavour will be to ensure the welfare of animals committed to my care'. The same duty must apply to farmers and stockpeople. The welfare state of an individual animal is conventionally defined as satisfactory by scientists and legislators when the animal is able to cope with the challenges of its environment and so, without difficulty, meet its physiological and behavioural needs (Broom and Johnson 1993). An individual animal is best able to meet its needs when the environment to which it is exposed is appropriate to the physiological and behavioural repertoire that it has acquired mainly through evolution but also through adaptation during individual development. At this early stage of the argument, good health may be considered as just one category of physiological need so that overall good welfare, or a state of well-being, may be defined with exquisite simplicity by the words 'healthy and happy' (Webster 2005).

The primary function of farm animals is the production of goods (e.g. food) and/or services (e.g. traction). There is a producer's argument which says that productivity is a good and sufficient measure of animal health and welfare, on the basis that if the animals were not healthy and happy they would not grow/

milk/lay so well. While it is possible to marshal some support for this argument it is very far from the whole truth. Here are two illustrations where it fails.

- An individual laying hen in a barren cage will continue to produce approximately one egg per day despite health problems of osteoporosis and bone fractures, and welfare problems associated with denial of most behavioural needs.
- An intensive, densely stocked unit rearing a population of broiler chickens to slaughter in 40 days is likely to be more economically productive (when the birds are sold at the same price/kg) than one that gives the birds more space and restricts feeding to achieve slaughter weight at 70 days, despite the fact that more *individual birds* in the former population will show mortality (e.g. from cardiac failure) and welfare problems associated with the pain of lameness.

Within the European Union, farm animals have, since the Treaty of Amsterdam in 1997, been classified not as commodities but as sentient creatures. This has generated a new basis for legislation that recognises that it is not enough to provide farm animals with appropriate feed, shelter and protection from disease, we should also consider how they feel. To do this we need to explore the nature of sentience itself.

Sentience and suffering

A sentient animal is a feeling animal, where the word feeling implies much more than simply responding to sensation. A frog with its head removed but its spinal cord intact will respond to a nociceptive[1] stimulus to its foot (a pinch) by withdrawing its leg. A sentient, conscious rat will respond similarly to a similar nociceptive stimulus (an electric shock) from the floor of its cage. If these shocks are repeated, the rat will learn to associate them not only with the physical sensation of pain but also an emotional sense of distress. The animal will be motivated by these sensations to behaviour designed to avoid the source of the shock. If it succeeds, it will learn that it can cope and will feel better. If it fails it may develop signs of extreme anxiety or profound depression (learned helplessness): i.e. it will feel progressively worse.

Animal sentience may be defined as 'feelings that matter' (Webster 2005). Marian Dawkins (1990) has pioneered the study of motivation in animals by seeking to measure how hard animals will work to achieve, or avoid, a resource or stimulus that makes them feel good, or bad. This recognises that the behaviour of sentient animals is motivated by the emotional need to seek satisfaction and avoid suffering. Many of these emotions are associated with primitive sensations such as hunger, thirst, pain and fear. There is an increasing body of evidence to show that farm animals also experience (and are motivated by) 'higher' sensations within the spectra of confidence–anxiety, hope–depression, friendship–grief (Webster 2011a). These things may expand the nature of their

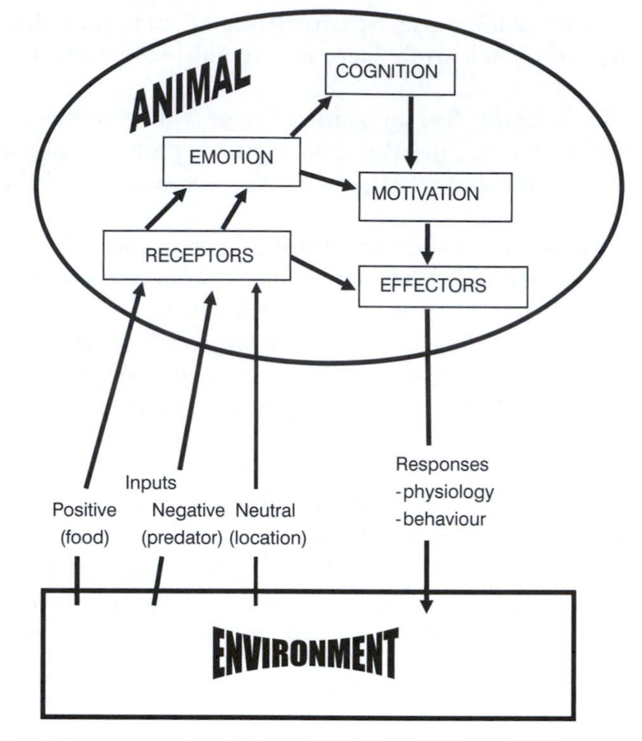

Figure 3.1 Sentience: an emotional view of life (from Webster 2005)

sentience but we should not underestimate the emotional stress caused by such things as hunger and pain. They may be primitive but this does not make them any less intense.

Figure 3.1 illustrates how sentient animals perceive their environment and how this motivates their behaviour (Webster 2005). The model is simple but soundly based in both experimental psychology and neurobiology. The 'control centres' in the central nervous system (CNS) receive information from the external and internal environment. Much information, e.g. our perception of how we stand and move in space, is processed at a subconscious level. Having learned to walk, we are able to control our limbs without recourse to thought or emotion. However, any stimulus that calls for a conscious decision as to action must involve some degree of interpretation. Motivation scientists observing the response of sentient animals may define a stimulus as positive, aversive, or neutral. In simpler words, the animal when presented with the stimulus will feel good, bad or indifferent. This is an emotional (i.e. sentient) response to the stimulus. The sentient animal (within which category we must include *Homo sapiens*) may or may not also interpret the incoming information in a cognitive fashion, i.e. apply reason. However they, and we, are usually and most powerfully motivated by how we feel.

To illustrate this point, consider the primitive sensation of hunger. The centres within the CNS responsible for control of appetite and satiety will respond to a variety of internal and external stimuli: low blood glucose, the sight or smell of food, or a conditioning stimulus (e.g. the bell that preceded the meal for Pavlov's dogs). All this information will be integrated in the form of an emotion. If the animal feels hungry it will be motivated to seek food. If a good meal arrives, it will feel pleasure. If no food is available, it will feel bad. This psychological concept of mind makes a clear distinction between the reception, categorisation and interpretation of incoming stimuli. Although it may appear abstract it has received strong support from neurobiology. Keith Kendrick (1998), for example, has made recordings from single neurones within the brains of sheep presented with external stimuli, or photographic images of external stimuli. A wide range of images (e.g. sacks of grains, bales of hay) trigger signals in a family of neurones that convey the generic information 'food'. A second set of stimuli or images, e.g. dogs and men, form another generic category of information that we may call 'predator'. The information 'food' then proceeds to a second processing centre where it stimulates a family of neurones that transmit a positive emotion (good). The information 'predator' passes to another centre that transmits the negative emotion (bad). However, if the sheep is now presented with a picture of a human carrying a sack of food, two categories of information (food and predator) are passed to the emotion centre, evaluated together and in this case passed on as a single, unconfused emotional message, namely 'good'.

The animal's decision as to how (or indeed whether) to respond is therefore determined by how it feels at the time, good or bad. Moreover, in a sentient animal (as distinct from, perhaps, an insect) the interpretation of information as good or bad is not a simple yes/no decision. The intensity of its feelings will vary. It will, for example, feel more or less hungry, more or less afraid, and this will determine the strength of its motivation to respond in positive or negative fashion. Thus by studying the strength of motivation of an animal to seek or avoid the feelings it associates with certain sensations and experiences, we can obtain a measure of how much these feelings matter.

Having acknowledged that sentience in animals implies feelings that matter, the second step is to acknowledge that sentient animals do not just live in the present. The sentient animal perceives, evaluates and interprets incoming sensation, decides how much it matters and makes a measured response. Having acted, the animal then assesses, emotionally and possibly cognitively, the effectiveness of its response. If it judges that it has achieved an effective response to a potentially threatening stimulus then it is likely to feel better when a similar event occurs in the future. If it judges that its response was ineffective, or if it was prevented by environmental or other constraints from behaving in a way designed to improve how it feels, then it is likely to feel worse.

The fact that the emotional response of animals to stimuli is governed by their past experience carries obvious survival advantages in a challenging environment. The importance of sentience in term of evolutionary fitness was recognised by Charles Darwin. The interpretation of past experience is equally important to a domestic animal since it is a key indicator of the animal's success, or otherwise, in coping with stress. An animal that has no sense of pain or fear, for itself or its offspring, is at a profound disadvantage in the struggle for existence. So too is an animal that cannot remember what gave rise to pain or fear in the past and how well or badly it coped. To illustrate this point, consider the difference between fear and anxiety. Fear is an emotional response to a perceived threat that acts as a powerful motivator to action designed, where possible, to evade that threat. It is also an educational experience since the memory of previous threats, the action taken in response to those threats and the consequences thereof ('was it less bad than I feared or worse?') will obviously affect how the animal feels next time around. If it coped, it is likely to habituate to the stimulus; it will be less fearful next time. If it failed to cope, then it may develop a chronic sense of anxiety whether or not the original threat reappears.

Stress and suffering are not the same. Animals are equipped to respond and adapt to challenge in circumstances that permit them to make an effective response. These emotions are normal and have evolved as key elements for survival. An animal is likely to suffer when it fails to cope (or has extreme difficulty in coping) with stress:

- because the stress itself is too severe, too complex or too prolonged (e.g. a dairy cow worn out by the sustained complex stresses of metabolic overload and chronic pain from lameness);
- because the animal is prevented from taking the constructive action it feels necessary to relieve the stress (e.g. a sow in the extreme confinement of an individual pregnancy stall).

Husbandry and welfare

The elements of good welfare, and the husbandry provisions necessary to ensure (or at least promote) good welfare, are encapsulated in the Five Freedoms as set out by the UK Farm Animal Welfare Council (FAWC; see Dawkins 1990). The Five Freedoms are accompanied by Five Provisions, a brief description of the husbandry (resources and management) necessary to promote each principle of good welfare (Box 3.1).

A more recent approach to categorisation of the elements of good welfare according to four principles and twelve criteria has been proposed as a result of the pan-European Welfare Quality® project (Botreau *et al.* 2007). The four WQ principles are essentially the same as the Five Freedoms. In each case they describe paradigms for an ideal welfare state; something to which those who care for the animals should aspire but can seldom hope to achieve. These paradigms of good welfare are also intangible. I can say that I am hungry or in pain, and give some indication of the intensity of these sensations, but how can I be sure how hungry you are – still less how hungry is this chicken or fish? The twelve welfare criteria linked to the four principles constitute a first step towards defining a series of tangible observations and measurements that can be used to gather information to help build up a picture of each of the welfare categories ('Freedoms' or principles) (Table 3.1).

Box 3.1 The Five Freedoms and Provisions

- *Freedom from thirst, hunger and malnutrition* – by ready access to fresh water and a diet to maintain full health and vigour.
- *Freedom from discomfort* – by providing a suitable environment including shelter and a comfortable resting area.
- *Freedom from pain, injury and disease* – by prevention or rapid diagnosis and treatment.
- *Freedom from fear and distress* – by ensuring conditions that avoid mental suffering.
- *Freedom to express normal behaviour* – by providing sufficient space, proper facilities and company of the animal's own kind.

Table 3.1 Categories of good welfare as described by the FAWC 'Five Freedoms' and the Welfare Quality® principles

Five Freedoms	*Welfare principles*	*Welfare criteria*
Freedom from hunger and thirst	Good feeding	Absence of prolonged hunger Absence of prolonged thirst
Freedom from thermal and physical discomfort	Good housing	Comfort around resting Thermal comfort Ease of movement
Freedom from pain, injury and disease	Good health	Absence of injuries Absence of disease Absence of pain induced by management procedures
Freedom from fear and stress Freedom to exhibit normal behaviour	Appropriate behaviour	Expression of social behaviours Expression of other behaviours Good human–animal relationship Positive emotional state

Problem analysis

The welfare of an animal is defined as satisfactory when it is able to cope within the environment to which it is exposed. The environment, defined in its broadest sense, presents a continuous, dynamic, ever-changing supply of physical and social information to the animal with regard to its comfort and security. It also presents materials essential to satisfaction and survival in the form of feed and water and it presents challenges to health and welfare in the form of infectious agents, parasites and predators. These are characterised as *inputs* in Figure 3.2. The complex interactions between all these inputs create the *scenario* with which the animal must cope. The simultaneous impact of all the factors that constitute a particular scenario will have *consequences* for its welfare (its ability to cope), as defined by:

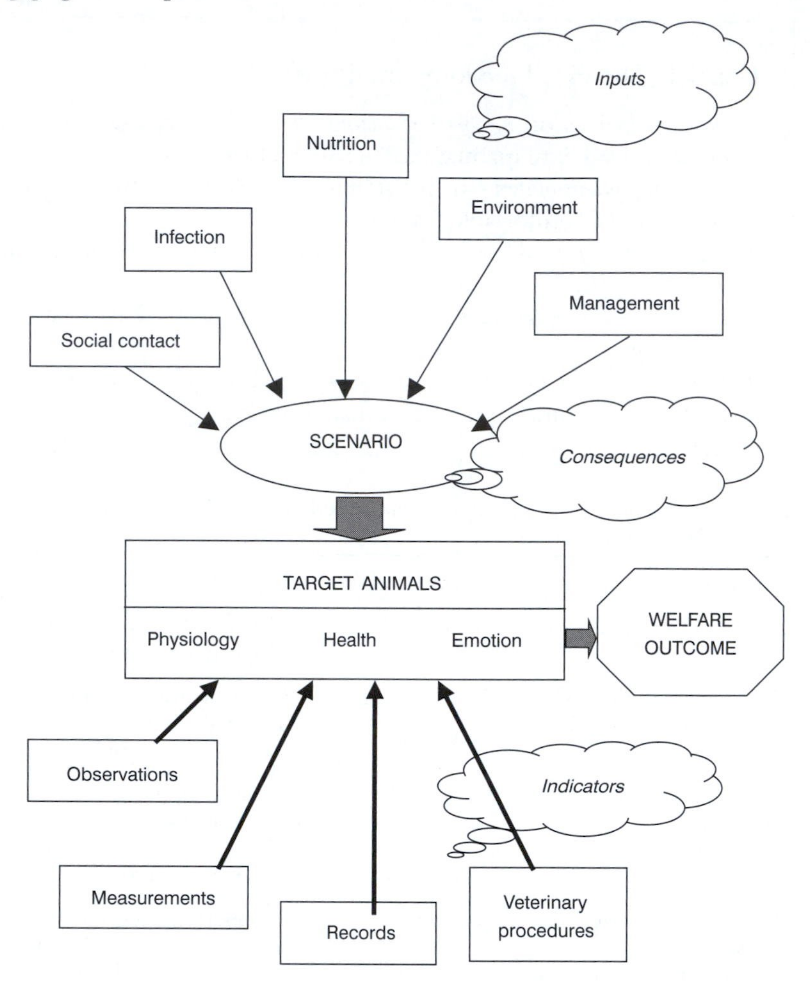

Figure 3.2 Animal welfare and the environment: inputs, consequences and measures

- *physiological state:* the ability to meet e.g. metabolic requirements and maintain homeothermy;
- *health status:* the ability to avoid e.g. injury and resist challenge from pathogens and parasites;
- *emotional status:* the ability to avoid e.g. pain and fear, and maintain a positive emotional state.

Animals in the wild are usually able to modify the scenario to which they are exposed by making choices as to physical habitat, feed intake, social and antisocial interactions with other animals. Moreover wild animals are normally adapted to their habitats because they have acquired by evolution the appropriate physiological and behavioural tools. Domestic animals that have been bred for extensive pastoral systems, whether in the hills of Scotland or the steppes of central Asia, have also been selected for traits designed to help them cope. Moreover (and this is extremely important), although these environments can, from time to time, be extremely challenging, the animals have considerable freedom of choice; the opportunity to make decisions that, *in their opinion*, constitutes the best coping strategy.

In intensive livestock systems, the environment (i.e. nutrition, housing, exposure to other animals and to infection) is determined by the farmer. Clearly, in the most intensive and efficient production systems (e.g. broiler chickens) there has been a large investment of science, technology and experience to create environments that appear to be optimal, when assessed in terms of the health and productivity of the population – although, as indicated already, this is not necessarily optimal for the welfare of each individual. Thus, when considering the scenario experienced by a confined farm animal fed a ration prepared by the farmer or feed company, it is necessary also to take into account the resources (e.g. housing design) and management (e.g. feeding regimes, health control programmes, Figure 3.2). While the aim is, of course, to optimise nutrition, housing and hygiene, at least in terms of physical welfare (physiology and health), the decisions are made by the producer, not the animal. If the conditions are perceived as unsatisfactory by the animal (especially in regard to its behavioural needs) then there is little that the animal can do to modify the scenario. One of the greatest criticisms of intensive livestock systems is that they seriously restrict the animals' capacity to make a constructive contribution to the quality of their own existence.

The ability of an animal to cope with a given scenario is determined by its *phenotype*, which is the anatomical, physiological and behavioural expression of its genotype. As indicated already, in wild animals and semi-domesticated pastoral animals, the phenotype is likely to have adapted to the environment through natural selection. In farm animals, however, genetic selection has been for traits selected to meet the needs of the producers: growth rate, carcass quality, milk production, feed conversion efficiency, and there is no reason to assume that these are positively correlated with traits consistent with robust health and welfare. Indeed, there is considerable evidence that excessive emphasis on production traits has led to serious disorders of health and welfare, of which painful leg disorders in broiler chickens and infertility in dairy cattle leading to premature culling are two conspicuous examples. Jacky Turner (2010) has produced a scholarly and comprehensive review of this problem.

The overall welfare outcome for the animal ('how fit is it and how does it feel?') is the consequence of the impact of all environmental factors that define a given scenario on its physical and emotional state, which is itself a function of its phenotype and past experience. However this is not something that one can measure directly. In order to analyse the welfare consequences of a stated scenario for an individual animal, or the consequence of a particular production system on a population of animals, it is necessary to make an appropriate selection of welfare *indicators* (Figure 3.2). These indicators may be categorised as:

- *animal-based observations and measurements* that can be made by a farmer, veterinarian or external assessor during inspection of the animal or farm unit (e.g. body condition, locomotion, injuries);
- *animal-based records* (e.g. mortality, fertility, treatment for infectious disease);
- *veterinary procedures* (e.g. information from blood samples and other procedures that must be made by a veterinary surgeon or other qualified individual).

These indicators need to be robust; i.e. measurable and recorded with a reasonable degree of accuracy and repeatability. From these indicators one can make an assessment of welfare according to one or more of the criteria listed in Table 3.1: e.g. absence of thirst, thermal comfort, absence of pain. In most cases this level of assessment should prove adequate as a basis for management support in relation to a specific problem at farm level or to ensure that there are no grounds for action by reason of specific failures to comply with welfare legislation. For purposes of quality assurance, it is scientifically acceptable to integrate assessment of welfare outcomes as far as the WQ four principles (or the Five Freedoms). The Welfare Quality® protocols (2009), which are primarily intended for use by quality assurance inspectors, rank farms according to each of the four principles as unclassified, good, enhanced or excellent.

For most purposes it is seldom necessary or indeed desirable to attempt a science-based integration of welfare outcomes that goes beyond these four (or five) categories. The WQ approach to passing judgement with regard to overall welfare state is (arbitrarily) based on the ranking for the four principles. To achieve an overall rating of *Excellent*, a farm must rate no less than enhanced for all principles, and excellent for two. To rank as *Enhanced* it must rate no less than acceptable for all four, enhanced for two.

The final assessment as to the acceptability or quality of a particular farm or a particular husbandry practice becomes a value (or political) judgement. Politicians always claim that their decisions in regard to animal welfare are based on the scientific evidence. However it is not possible to decide, on the basis of science alone, whether (e.g.) freedom from hunger, discomfort and disease adequately compensates for denial of freedom to exhibit normal behaviour. A classic illustration of this is provided by differences in legislation regarding the keeping of pregnant sows in individual stalls. The European Community, having reviewed the scientific evidence, concluded that the sow stall was unacceptable and have imposed a ban effective from 2013. Legislators in Australia and the USA examined the same scientific evidence and concluded that the practice was acceptable. The differing conclusions arose from the fact that the different groups gave different values to the different principles – essentially the relative importance to the animal of the physical and emotional elements of welfare. However, the situation is not static. I am happy to say that, at the time of going to press, the sow stall has now been banned in Australia and several states of the USA. The science has not changed. The driver for this and similar changes in animal welfare legislation has been consumer pressure which is progressively overcoming producer pressure.

To summarise this section: analysis of the impact of the environment on the health and welfare of animals requires identification, selection and, where possible, measurement, of factors that make up the environmental scenario. It is then necessary to select robust and appropriate indicators from which to create an assessment of welfare state according to defined welfare principles (e.g. freedom from hunger, pain, fear). The approach is generic. Later I shall show how it may be applied to specific cases.

The environment, animal welfare and disease

The issue of farm animal welfare and disease needs to be viewed from both directions: disease as a threat to welfare, and poor welfare as a precursor of disease. Table 3.1 defines one of the central principles of good welfare as 'good health', or 'freedom from pain, injury and disease'. Disease, by this definition, covers any departure from good health. For farm animals the most important disease categories include infections with pathogenic microorganisms, infestation with parasites, injuries and locomotor disorders, and disorders of digestion, metabolism and reproduction. It is also important to acknowledge the separate (but overlapping) category of *production diseases*: i.e. diseases attributable wholly, or in part, to conditions of feeding, housing and hygiene imposed by the management system deemed necessary for commercial production. These may include infectious conditions such as mastitis, metabolic conditions such as ketosis, and locomotor disorders presenting as lameness. The individual conditions that make up these categories are disease are legion, but the approach to their management can be considered by category.

I repeat: the health and welfare of a sentient animal are defined by its physiological and emotional state as it seeks to cope with challenge. Challenges ('stressors'), which may be physical (e.g. cold), emotional (e.g. fear) or biological (e.g. infectious organisms), induce a stress response defined by three phases: alarm, adaptation and exhaustion. The *alarm* phase, which involves the hypothalamus/pituitary/adrenal (HPA) axis, is brief and non-specific. It may induce immuno-suppression and increase susceptibility to infectious disease but in most cases it is the lasting effects of exposure to prolonged or repeated stress that are likely to be more important. The second phase of the response is *adaption,* which may be partial or complete. Faced by a physiological stressor such as cold, a cow or chicken must increase heat production in order to maintain homeothermy. This imposes a metabolic cost. In time it can adapt through mechanisms to improve insulation (e.g. by growing a longer coat). If this is sufficient to eliminate the metabolic cost, adaptation is complete; if not, the cost will be reduced but not eliminated. At the *exhaustion* stage, where cold is exacerbated by loss of body condition, the cost will increase and may prove fatal. In the adaptation phase the response tends to be specific to the stressor (e.g. cold) and the severity of the stress is measured by the magnitude of the physiological response (increase in heat production), not the stressor (e.g. air temperature). The capacity to cope with other, simultaneous stressors may or may not be impaired.

Stressors, whether from the physical, microbiological or social environment, provoke emotional as well as physical responses. These affect how the animals feel and act at the time and the memory of these feelings determines their attitude, expectation and response to the presence or threat of their recurrence. If they learn through experience that they can cope, they adapt or habituate. If they learn that they cannot cope, or have extreme difficulty in coping, then stress may proceed to suffering. Moreover, increasing disturbance to their physiological and emotional equilibrium is likely to increase their susceptibility to other stressors such as infectious organisms. Here again, however, the severity of the response to physical and emotional stressors can only be assessed in terms of measured indicators of the animal's response.

Interactions between the environment (as a source of stressors), infection, disease and animal welfare are illustrated in Figure 3.3. When a susceptible population of animals is exposed to a highly infectious pathogen such as foot and mouth virus (FMV), exposure to infection leads almost inevitably to disease. This is illustrated by the curved arrow in Figure 3.3. However, the major infectious diseases of farm animals are usually controlled by a policy of vaccination or eradication. Most infectious diseases of most farm animals are endemic, where infection may be considered the natural state and disease as an expression of failure to maintain the balance between potential pathogen, host and the environment. This applies both to the endemic respiratory and enteric infections that manifest as production diseases in intensively farmed animals and to the parasitic diseases of pastoral animals owned by subsistence farmers.

Figure 3.3 identifies six factors involved in the pathogenesis of such endemic diseases: pathogen virulence and environmental challenge, host welfare and adaptive state, immunity and infection. Each is illustrated by a wedge or triangle to indicate that each state is a variable. The state of adaptation will be low when the cost of coping with stressors is high or the animal approaches exhaustion. Adaptation state is high when the animal has made effective physiological and emotional responses sufficient to minimise or eliminate the effects of the stressors. Variation in adaptive state will influence the progression from pathogen exposure to disease or resistance largely, though not exclusively, through impact on the immune system. When the immune system is competent, exposure to infection is likely to progress to a state of resistance. When the immune system is suppressed then infection is more likely to progress to clinical disease. These consequences will then feed back to the animal's welfare. Disease will impair welfare, resistance preserve or enhance it.

Many fundamental scientists, deeply involved in the study of infection and immunity at the cellular or molecular level, may dismiss this model as simplistic. I contend that it is quite the reverse, since it recognises that any study of farm

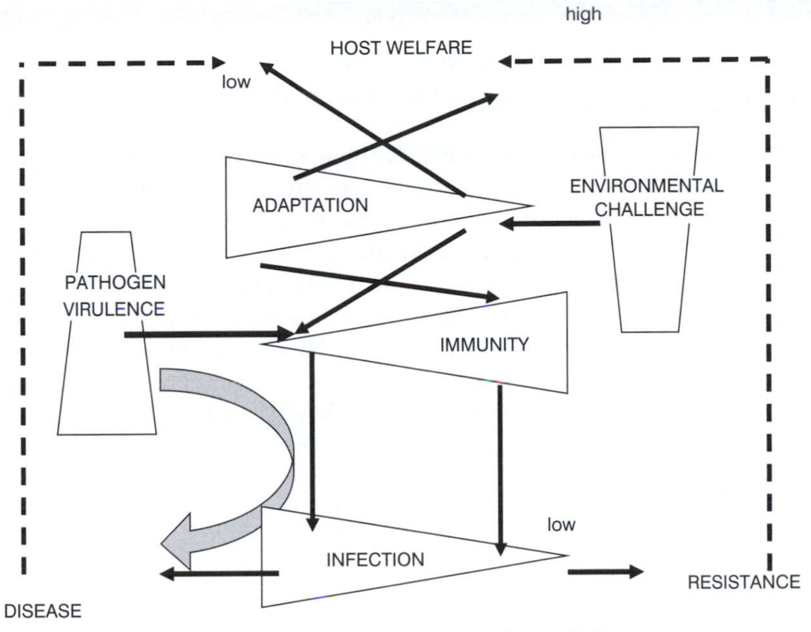

Figure 3.3 Interactions between the environment, welfare and disease

animal welfare and the interface with disease needs to take all these factors into account. In particular, it makes the point that study of interactions between host, pathogens and the environment is a multidisciplinary affair. It requires teams containing individuals with reductionist skills *and* those who can comprehend the broader picture.

Infectious diseases

Box 3.2 describes the categories of infectious disease in outline. The first categories are the epidemic diseases, whether caused by indigenous or exotic pathogens. The distinction is important to a farmer or veterinarian operating in a particular country or region. However it has no meaning when considered in an international context. Typically, these pathogens are viruses with a relatively high virulence, which favours the development of clinical symptoms (e.g. fever, sneezing) and the rate of transmission from infected carriers to other susceptible (non-immune) individuals. These properties also carry the seeds of their own decline, either because they kill the animals and the pathogen with them, or because they lead to a rapid build-up of immune individuals who block the process of transmission. In many cases a viral epidemic simply dies out when the number of carrier animals becomes so small relative to the number of immune individuals that the pathogen cannot survive. In other cases, the epidemic appears to disappear, because the proportion of immune animals is very high, but the pathogen remains at large, ready to strike again when the immune status of the population falls.

Box 3.2 Categories of infectious disease

Epidemic diseases (indigenous)

Infections resulting from exposure of a population of animals to a pathogen that is novel to that population, though present in the country or region. The rate of new infections will be defined by the rate of transmission of infection among non-immune individuals. It rises while the number of infected carriers increases and the number of immune individuals remains relatively low. As the number of immune individuals rises as a result of recovery from infection, or vaccination, the proportion of susceptible individuals reaches a peak then declines. If the number of infected carriers falls below a critical number, the epidemic will be over.

Principal hazards

Animal transport, vectors (insects, wildlife, humans), failures of biosecurity, failure to vaccinate.

Prevention and control

Biosecurity (including creation of specific-pathogen free facilities), hygiene, vaccination.

Epidemic diseases (exotic)

Epidemic diseases associated with pathogens not normally present in a country or region.

Principal hazards

International transport of animals and animal-based feedstuffs, insect and bird vectors.

Prevention and control

Eradication by slaughter when it is possible to control transmission (e.g. foot and mouth virus), or vaccination, when it is not (e.g. Bluetongue virus).

Endemic diseases

Diseases associated with continuous exposure to indigenous organisms that sustain their capacity to infect populations of animals through their capacity to survive in the environment or in relatively healthy carrier animals (e.g. enzootic pneumonia in pigs and calves).

Principal hazards

High pathogen challenge associated with high stocking densities and poor hygiene, depressed immune status, introducing non-immune animals to populations of relatively healthy carriers.

Prevention and control

Housing and hygiene, stockmanship.

Zoonotic diseases

Infections transmissable between animals and man.

Endemic

Conditions (e.g. salmonellosis, campylobacter) where there is a reservoir of infection in the (farm) environment and population of relatively healthy carrier animals.

Prevention and control

Elimination of carriers, hygiene, vaccination.

Epidemic

Diseases caused by (e.g.) influenza viruses (H5N1, H1N1) that can infect both humans and farm animals (pigs and poultry).

Prevention and control

Vaccination, biosecurity.

Transmissable encephalopathies

BSE/nvCJD: Diseases caused by prions (abnormal proteins capable of self-replication that can be transmitted between animals, or animals and man, by consumption of 'infected' materials (e.g. central nervous system tissue).

Major epidemic diseases of farm animals include foot and mouth disease (FMV) affecting cloven hoofed animals (cattle, pigs and sheep), several varieties of swine fever, Newcastle disease (or fowl pest) and bluetongue in sheep. All these diseases are highly infectious to non-immune animals, have a relatively high mortality rate and severely affect the productivity of those animals that do not die. Thus they are of major economic consequence. In Europe these are notifiable diseases and control measures are determined by legislation. In Great Britain, FMV is treated as an exotic disease, to be eliminated from the island as quickly as possible. The disease has occasionally entered GB by airborne transmission, but the more common origin has been infected imported meat fed to pigs in improperly treated swill. The policy is to slaughter all affected animals and all those deemed to be in close enough proximity to be at risk. This, combined with strict control of animal movements and strict attention to hygiene, does eliminate the disease eventually. In other parts of the world FMV is controlled through a policy of mass vaccination, or slaughter of animals in the immediate vicinity accompanied by the creation of a firewall of vaccinated animals around the infected area.

The first justification for the British policy of mass slaughter is epidemiological. We are an island nation which should make it relatively easy to keep the virus out. The second justification is politico-economic and driven more by dogma than by logic: namely that if Britain is perceived to be not entirely free of FMV this will seriously reduce the value of exports of cattle and the products of cattle. At the beginning of the last (2001) outbreak of FMV in the UK, the European Community imposed a temporary ban on the export from the UK of livestock, meat and animal products. Annual income at the time from these sales was estimated at about £500 million. The direct cost of the FMV policy to government (i.e. the taxpayer) was estimated as £2.1 billion, to which may be added at least another £1 billion in indirect costs due to losses of tourism and rural industry (*Private Eye* 2002). The economic argument for a rigid policy of eradication by slaughter appears very weak, even before one begins to consider the cost in suffering to the animals and the farming population.

There has been another greater and more lasting economical and environmental cost arising from this FMV control policy. Following the 2001 FMV epidemic, the European Commission prohibited the feeding of catering waste to farm animals (Animal By-products Regulation (EC) No. 1774/2002). This included food wastes from vegetarian restaurants. This denied the pig industry the opportunity to farm pigs in complementary mode as scavengers of the approximately 40 per cent of food that we throw away. In a society less paralysed by an innumerate concept of risk and paranoid fear of litigation, every chain of supermarkets would operate its own pig units, wherein the pigs were fed on the food we discard because it doesn't look pretty, has the wrong shape or has passed its sell-by date. However, there is now a glimmer of light. One feed company in UK has been granted dispensation to produce sterilised pig feed based on supermarket waste.

Bluetongue, a virus which principally affects sheep and has similar effects to FMV, has recently spread north and west across Europe. This disease is primarily transmitted by an insect vector (the *Culicoides* midge), which is impossible to exclude when the temperature and humidity are right (for the midge) and the wind is in the wrong direction (for the sheep). The advantage of this form of transmission is that not even the most ruthless government veterinarian could conclude upon a policy of eradication. An effective policy of vaccination has been applied and, in general, everyone is satisfied.

Endemic diseases

Endemic diseases are associated with continuous exposure to indigenous organisms that sustain their capacity to infect populations of animals through their capacity to survive in the environment or in relatively healthy carrier animals. For purposes of characterisation, we may identify two types:

- endemic infections of intensively reared animals housed in close proximity to their conspecifics;
- endemic infections of free-ranging animals exposed to infection from conspecifics or wildlife.

Some of the most serious problems of endemic disease are seen in intensive pig production systems. These include enzootic pneumonia, post-weaning enteritis and PMWS (porcine multiple wasting disease). Enzootic pneumonia in fattening pigs has a low mortality rate but, in confinement systems, can significantly impair growth rate and feed conversion efficiency. All these may be considered as production diseases (see below).

A troublesome example of the second category of endemic disease (in the UK) is bovine tuberculosis (bTB). Despite a rigorous control programme involving regular tuberculin testing, slaughter of animals showing positive reactions and severe restrictions on movement of all animals on farms with reactors, it has not been possible to establish control, especially in the dairy cow grazing areas of south-west England, Wales and Ireland. One of the reasons for this is that bTB is endemic in wildlife popultations, especially badgers, whose population numbers have greatly increased in recent years. Slaughter of the badger population would reduce the risk of bTB infection in cattle. However, it would have to be on a very large scale as local eradication is usually followed by repopulation with new badgers who are also likely to be infected. It also poses a stack of socio-economic and ethical questions. 'Is it right to slaughter badgers (and other wildlife, including red deer) on a massive scale to improve the health and the economic efficiency of the dairy industry?' More pragmatically, 'will the public buy it?' The UK government has faced these dilemmas by commissioning an endless succession of reviews and research programmes that have served their (the government's) purpose by postponing the time for decision. Finally they appear to be moving towards a strategy based on selective slaughter in areas of high infection supported by vaccination in surrounding areas. Existing vaccines are quite good for this purpose and vaccines under development should be better. The policy of vaccinating the badgers rather than the cows is a sound one. To guarantee freedom from infection in a herd or region, a vaccine for cows would need to be 100 per cent effective and immunologically distinstinguishable from natural infection with *M. bovis.* However a badger vaccine that is only 70 per cent effective should be sufficient to achieve a major reduction in TB prevalence within the badger population, thereby greatly reducing the risk of cross-infection to cattle. The TB problem is but one example of the risks involved in allowing food animals and wild animals to share the same habitat. However, it is entirely natural that different species should share the same habitat and it is entirely natural that they should be exposed to a wide range of pathogens and parasites. Indeed, mixed grazing, whether of cattle and sheep in the UK, or ruminants and equids on the African plains, is essential to maintain the equilibrium between host, parasite and environment as illustrated in Figure 3.3. It is trite but true to say that endemic infections present problems when the balance of nature is disturbed and the pathogen/parasite challenge exceeds the coping capacity of the host animal, e.g.:

- overstocking pasture land, especially with a single species;
- overstocking housed animals, especially when hygiene and ventilation are inadequate.

These further examples of the problems arising from the premise that the sole purpose of animal husbandry is to provide the commodity of cheap food leads to a return of my big theme, namely the concept of 'planet husbandry': multipurpose management of land for food, amenity, carbon sequestration, wildlife conservation etc., whereby the income from the land does not depend on any single commodity (e.g. food) and the health of the living environment is preserved for all. It is also valid to point out that a policy that prohibits the selective killing of a single protected species at the top of the food chain (e.g. the badger) will inevitably contribute to the problems of overpopulation (e.g. bTB). These problems will be considered further in Chapter 8.

Zoonotic diseases

Zoonotic diseases, infections transmissable between animals and humans, present us with some of the most serious risks associated with the rearing of animals for food. We may identify three categories.

- Endemic bacterial infections of food animals, e.g. *campylobacter* and *salmonella*: transmissable to humans via contaminated meat, eggs, unpasteurised milk or from direct contact with animal excreta.
- Epidemic viral infections, e.g. H5N1 and H1N1 influenza: caused by highly mutable viruses that can infect several species, typically pigs, poultry and humans.
- Transmissable encephalopathies, especially bovine spongiform encephalopathy (BSE): transmissable to humans as new variant Creuzfeld Jacob disease (nvCJD).

The most common zoonoses are those caused by food poisoning micro-organisms, of which the most important are *campylobacter* and *salmonella* species. Infection with *campylobacter,* which can cause severe, but usually short-lasting diarrhoea in humans, most commonly results from eating undercooked, contaminated poultry meat. The current incidence of *campylobacter jejuni* in raw chicken carcasses in Europe is over 50 per cent (Corry and Atabay 2001), which means that the disease is not under control. The organism inhabits the chicken gut and is relatively harmless to the birds. However under conditions of stress (e.g. transport prior to slaughter) more organisms are shed from the gut and also absorbed into the systemic circulation, increasing the risk that the organism will be present not only on the surface but also deep in muscle. In this case, classic approaches to disease control, including purchase of chicks from *campylobacter*-free breeding stock and strict attention to manure disposal and disinfection between batches of birds are ineffective, since the infection appears to be carried in the air and (especially) drinking water. The prevalence would seem to be similar in intensive and less intensive systems of broiler production. However the industrialised slaughter of birds (which have been stressed before slaughter) undoubtedly increases cross-contamination between carcasses. While there is, at present, no effective method for control or elimination of *campylobacter* in poultry flocks, management of the risk of food poisoning is in our hands (kitchen hygiene and proper cooking).

There are many, many strains of *salmonella* species, of which the most virulent for humans is *S. typhi,* the organism responsible for typhoid fever. Two of the most common zoonotic strains are *S. typhimurium* and *S. dublin*. These organisms can cause severe disease and death in animals, especially young calves, so famers are powerfully motivated to keep them under control. Another organism, *S. enteritidis,* endemic in flocks of apparently healthy laying hens, is the major source of food poisoning from eggs. Here again, the most effective control is proper attention to food hygiene. One lightly boiled egg may not carry a sufficient infective dose. However, when live salmonella organisms are allowed to multiply in an egg salad or on a tray of processed meat, then the results can be severe. Blame for the current high incidence of human food poisoning cannot be laid primarily at the door of the intensive farm. The major risks occur beyond the farm gate: at the abattoir, in the processing plant, on the butcher's or supermarket shelf or, most especially, in the kitchen.

A secondary, and ultimately more serious problem is the increased appearance of antimicrobial resistance in *campylobacter* and *salmonella* species recovered from the faeces of cattle, pigs and poultry and food from these animals. There is no doubt that this is linked to the routine administration of antimicrobials to the food animals. Resistance to antibiotics such as tetracycline and ampicillin that have been in common use for years is now widespread (Bywater *et al.* 2004). Moreover there is increasing resistance to 'new' antimicrobials such as ciprofloxamin (fluoroquinolone) that is 'of critical importance to human medicine' (EFSA 2010). The routine use of antimicrobials to control endemic disease and improve feed conversion efficiency in intensive pig and poultry units has undoubtedly contributed to the problem of antibiotic resistance in humans, although less than the prescription and use of antibiotics in humans (doctors handing out unnecessary prescriptions and patients failing to complete the dose).

The European Council and Parliament banned the use of antibiotics as growth promoters in the food animals with effect from 2006. They state that antibiotics used in human medicine must not be included as food additives, 'while preserving animal health': a let-out clause that permits farmers to employ veterinarians to prescribe antibiotics on a more or less routine basis for allegedly 'therapeutic' purposes. There is a fine line between mass administration of therapeutic antibiotics to all animals to reduce the burden of infection and 'metaphylaxis': treatment of batches of animals when individuals show signs of clinical disease. This problem cannot be removed by legislation. The best solution is to develop systems of feeding, housing and hygiene that make routine antibiotic use unnecessary; which, in the current climate, equates to uncompetitively expensive.

The most important epidemic zoonoses are those caused by the family of influenza viruses that thrive among populations of pigs, poultry and people, most conspicuously within the village farms of South-east Asia. These viruses are made up of a central core of RNA and protein surrounded by an envelope of glycoproteins: haemagglutinin (H) that binds the virus to target cells and permits entry of the viral genome, and neuraminidase (N), which is involved in the release of progeny virus from infected cells. The most notorious strains are H5N1 (bird flu) and H1N1, responsible for the pandemics of Spanish flu in 1919 and swine flu in 2009. The popular names are misleading since the source is seldom exclusively the bird or the pig (still less the Spanish). The surface glycoproteins are highly prone to mutation and new or returning combinations acquire prominence almost every year, which is why public health programmes recommend the routine annual vaccination of people at high risk (e.g. the aged) with 'this year's' vaccine.

Intensification of pig and poultry production is *not* a risk factor for influenza. Indeed, the confinement of pigs and poultry within buildings positively ventilated by fans drawing air in through high-efficiency particle filters undoubtedly reduces the risk of the animals becoming infected from carriers such as wild birds. Nevertheless improved control of influenza in pigs and poultry would be a major step forward in terms of both animal and human health. This is an active area of research and there are some promising developments, such as the development of a vaccine effective against antigens common to all flu strains and the genetic engineering of birds resistant to flu. I shall discuss these further in Chapter 7.

The third class of zoonoses is the transmissable spongiform encephalopathies (TSEs). The most serious of these has been bovine spongiform encephalopathy (BSE or 'mad cow disease'). This was first identified in dairy cows in the UK in 1986 and it rapidly developed into an epidemic. It presented as a progressive deterioration of mental ability (awareness, behaviour, motor ability) associated with a degeneration of the functional protein structure of the brain. Similar diseases, including scrapie, transmissable between sheep, and sporadic Creuzfeld Jacob disease (CJD) in humans (rare and apparently not transmissable) had been known for years. The appearance of BSE in epidemic form was undoubtedly due to the inclusion of meat and bone meal of cattle origin in rations for dairy cows and their early-weaned calves, and exacerbated by a rendering process that failed to destroy the causative agents, prions – corrupted proteins existing in (especially) the nervous tissue of infected animals. Once this was recognised, an absolute ban on the feeding of any material of animal origin to cattle was imposed in 1989.

The first three deaths from new-variant CJD in humans occurred in 1995. The disease differed from typical sporadic CJD in that it presented in young people, and the dreadful progression of symptoms through dementia, hallucinations and paralysis to death was much more rapid. Laboratory tests confirmed that the prion responsible for nvCJD was identical to that causing BSE in cattle and further restrictions were imposed to prevent nervous tissue and other specified offals from cattle entering the food chain. The death rate from nvCJD peaked at 28 in 2000, the total number of deaths to 2011 was 171 but over the last four years the annual death rate has been below three. This has been a terrible lesson and there may well be more deaths in people infected before 1989 but not yet presenting clinical symptoms. However, the current risk of contracting CJD from eating meat is effectively zero, although the death rate in UK from sporadic CJD (not contracted from beef) remains at about 100 cases per year.[2]

Production diseases

Production diseases are those that arise directly from risks associated with the production system itself: failures in feeding, housing, hygiene and management, exacerbated in many cases by the breeding and selection of animals not fit for purpose. The main categories of production disease are summarised in Box 3.3. Disorders of digestion include rumen disorders like bloat and acidosis in dairy cows fed diets containing too much starch and too little fibre in an attempt to provide enough metabolizable energy (ME) to sustain the demands of lactation. Other disorders within this category include post-weaning diarrhoea in piglets. This is conventionally recognised as an endemic infection with organisms such as *Escherichia coli* that calls for prevention and cure through the routine administration of antibiotics. In fact, the primary cause is profound damage to the intestinal epithelium and normal commensal microbial population arising from too-soon weaning off sow's milk and too-early presentation of novel feeds to an intestine that is both enzymatically and immunologically immature and unable to cope. Opportunist organisms like *E. coli* then exploit the damage. A simple (though unpopular) solution is to delay weaning. The feed industry is more interested (and has had some success) in the use of feed additives to enhance digestibility. However they have not yet resolved the problems of allergic inflammatory reactions to feed antigens.

High-yielding dairy cows are at especially high risk from metabolic disorders. Acute conditions such as ketosis and post-parturient hypocalcaemia (milk fever) can be simply attributed, respectively, to the sudden increase in energy demand and calcium demand in early lactation. More chronic conditions such as loss of body condition leading to infertility are due to the cow failing to ingest

Box 3.3 Categories of production disease

1. Disorders of digestion and metabolism

Arising from the provision of feed that is inconsistent with healthy digestion (e.g rations for high-yielding dairy cows, early-weaned pigs, veal calves). Metabolic diseases arising from selection of animals for high production (e.g. dairy cows, laying hens).

Principal hazards

Overemphasis on high production, improper diet formulation.

Prevention and control

Modified diet formulation, modified breeding indexes to give less emphasis to production traits relative to traits related to robustness.

2. Reproductive disorders

Failure to conceive, periparturient injuries and infections.

Principal hazards

Loss of body condition in breeding females, failures in oestrus detection.

Prevention and control

Improved nutrition, improved oestrus detection, improved management at parturition and during lactation.

3. Infectious disorders

Diseases associated with exposure to endemic microorganisms but precipitated by inadequacies in breeding, housing, hygiene and management. (e.g. mastitis in cattle, post weaning diarrhoea, enzootic pneumonia and PMWS (porcine multiple wasting syndrome) in pigs.

Hazards

Inadequate housing and hygiene, overstocking, lack of biosecurity.

Prevention and control

'5-point' mastitis control plan (cattle), creation of specific-pathogen free rearing units (pigs).

4. Locomotor disorders

Lameness attributable to painful injuries and infections of the foot (e.g. dairy cattle), 'leg weakness' (e.g. broiler chickens), osteoporosis (laying hens).

Principal hazards

Inadequate housing, hygiene and routine foot care (cattle and sheep), unsatisfactory breeding policies, overstocking and improper feeding (poultry).

Prevention and control

Early diagnosis and treatment, (cattle and sheep), improved feeding strategies.

enough ME to meet the demands of lactation (Chapter 2). In all these cases, the problem can be resolved in part through better attention to diet formulation and feed presentation. However, one of the major risk factors has been the genetic selection by the breeders of sires and dams for milk production largely to the exclusion of other traits consistent with sustained good health and lifetime productivity; this was compounded by the enthusiasm of farmers for semen from bulls ranked high for production relative to those whose records show their progeny to be more robust. However, at last the high-input, high-output end of the international dairy industry has begun to acknowledge the errors of its ways and there is now a strong movement towards the breeding of dairy cattle on the basis of 'lifetime' production (in practice total yield and efficiency over 1,200 days from first calving) rather than peak production in first lactation.

At this time it is good to report the failure of BST (bovine somatotropin or growth hormone) produced synthetically and heavily promoted, initially within the USA, then internationally, on the basis that it could increase milk production by 15–20 per cent. Since the high genetic merit cow already had extreme difficulty in consuming and digesting enough nutrients to sustain 'natural' milk yield, and average life expectancy in the USA had fallen below three lactations, this never looked like a sustainable option. It was heavily promoted by Monsanto and others on the basis that farmers who didn't use it would go out of business – which is, of course, the morality of the protection racket. The European Commission, which differs from the USA in being more sensitive to public opinion and less to the agriculture lobby, banned BST, partly on valid grounds of animal welfare, partly on spurious grounds of human health but mainly in recognition of the fact that people neither wanted it or needed it. Reassuringly, in the USA, the use of BST has slowly withered away, thanks hardly at all to state legislation, almost entirely to the power of the people. A great victory for 'politics by other means'.

Some of the production diseases associated with infectious organisms have been discussed already. These include post-weaning diarrhoea, enzootic pneumonia and PMWS (porcine multiple wasting syndrome) in pigs, linked to disorders of feeding (too-early weaning), housing (overstocking, inadequate ventilation and poor hygiene) and management (failures in biosecurity). The other high-impact production disease associated with a multiplicity of microorganisms is mastitis in dairy cattle. The aetiology and control of mastitis are too complex to consider here. Very simply, one can identify two categories: contagious mastitis associated with failures of hygiene and management in the milking parlour, and environmental mastitis associated with faliures of hygiene in the cow house. Both types have a genetic component: selection for increased milk yield coselects for increased milk flow rate, and this increases the risk of bacteria entering the udder through the teat canal. Nevertheless the main causes of mastitis can be linked to environment and management, which means that they can be controlled by better attention to these things. Mastitis is undoubtedly a distressing condition for the cows, and *E. coli* environmental mastitis in early lactation can be fatal. It is also an expensive problem for the farmer due to costs of treatment and the milk that cannot go for sale. Consequently mastitis control on the best dairy farms tends to be very good. Thus, while it remains a major production disease, it does not rank high on my list of husbandry failures in urgent need of repair.

Dairy cow lameness, on the other hand, does. This, in my opinion, is the production disease that presents the most severe problems of welfare. The two most common categories of lameness in dairy cows are:

- damage to the horn of the cloven hoof (e.g. sole ulcer, white line disease), typically an outer (abaxial) claw on the hind foot;
- infection of the surrounding skin (e.g. digital dermatitis).

Lame cows are in pain, the severity of which can be assessed from observations of behaviour, ranging from a slight limp to extreme reluctance to move at all or put any weight on the affected foot. If more than one foot is badly affected the cow has a real problem. The evidence that pain is the problem (not some mechanical disorder) is overwhelming since the most effective remedies are those that involve pain relief (analgesic drugs, or relieving the affected claw from weight bearing by cementing a 'block' to the unaffected claw).

Most studies of lameness in commercial herds of high-yielding dairy cows in recent years have recorded a prevalence greater than 20 per cent, in conditions that range from year-round housing (zero grazing) in Europe and USA, to year-round at pasture in New Zealand. Simply stated, this means that approximately one-quarter of our commercial cows are in more or less severe pain. The risk factors for dairy cow lameness are many and varied, including:

- phenotype: size, conformation (especially legs and feet);
- physiology: structural changes in the foot linked to calving and early lactation;
- housing: cubicle design and bedding, floor type, slurry management;
- nutrition: inappropriate feeding of starchy concentrates;
- management: inadequacies of routine foot care (e.g. claw trimming), hygiene (antibacterial foot baths), failures to detect and treat cases at an early stage.

There is no doubt that general trends in the commercial dairy sector have contributed to the increasing prevalence of lameness. These include increased exposure to concrete and cows becoming too big for their cubicles. However the most serious risks are farm-specific. Recent evidence from Europe and USA reveals that the greatest risks on high-prevalence farms are due to failures of management: inadequate foot trimming, foot hygiene, and especially failure to diagnose and treat newly lame cows at an early stage. Improved attention to early treatment has been shown to reduce overall prevalence and practically eliminate the suffering caused by severe problems such as the sole ulcer (an appalling condition where the pedal bone has actually penetrated the sole of the foot).

Behaviour as an indicator of animal welfare

So far, we have considered consequences of husbandry provisions and husbandry decisions (breeding, feeding, housing and management) mainly in terms of the physical health and welfare of farm animals. Now let us consider the impact of husbandry on the emotional state of sentient animals as expressed in terms of their behaviour and motivation to behaviour. It is necessary to restate some basic principles. An elegant introduction to the application of scientific method to investigate behaviour as an indicator of animal welfare has been prepared by Christine Nicol (2011):

- Sentient animals are motivated by how they feel. Their behaviour (wherever possible) is directed by the need to avoid distress and promote a positive emotional state.
- An animal that learns that it is able to cope with environmental challenge through manifestation of normal behaviour is likely to achieve a positive emotional state. When the animal learns that it cannot cope, then its emotional state is likely to deteriorate and this may be expressed in forms of abnormal behaviour.
- Intensive production systems in which animals are kept in close confinement are likely to restrict their ability to address challenges to their emotional state through constructive behaviour designed to relieve stress. This increases the risk of development of abnormal behaviour.

Pain and disease

For sentient animals, including ourselves, pain is more than an unpleasant physical sensation; it also has a profound effect on our emotional state. Similarly disease presents more than just physical effects such as fever and inflammation, it also induces a sense of *malaise*, i.e. feeling ill. A pig or calf in acute pain (e.g. during castration) will vocalise and make vigorous attempts to escape. Animals in chronic pain (e.g. lame cows or chickens) will not just attempt to avoid contact and restrict movement at the affected site, they will also show behavioural changes indicative of a more general deterioration in welfare, e.g. reduced appetite, grooming and social behaviour. Chronic pain presents a particularly severe problem for animal welfare, since animals do not habituate (i.e. display a reduced response to a constant incoming stimulus). On the contrary, chronic pain leads to the phenomenon of hyperalgesia or 'wind-up'. In these circumstances, there is not only increased intensity of sensation at the site of injury, but hypersensitivity at unrelated sites so that everything that used to be just mildly uncomfortable now hurts a lot.

There are many other behavioural indicators of pain, the emotional impact of pain and the memory of pain. These include attempts to avoid exposure to events that have caused pain in the past (e.g. electric fences, handlers who wield sticks). There is also good scientific evidence that mammals (e.g. rats) and birds (poultry) in chronic pain from injury are able to select feed containing analgesic drugs. They learn within a few days that when they choose the feed that contains the analgesic, they feel better. They don't have to understand why it works but they accept that it does. In this regard their behaviour is more convincing than humans taking analgesics since there is no possibility of a placebo effect.

Fear and anxiety

Acute fear in the presence of real or imagined threat is a highly effective emotional response that acts as a powerful motivator to action designed, where possible, to evade that threat. It is also an educational experience since the memory of previous threats, the action taken in response to those threats and the consequences thereof ('was it less bad than I feared or worse?') will obviously affect how the animal feels next time around. Thus fear, like pain, is an essential part of sentience. The concept of fear as a stimulus to which animals may adapt, or fail to adapt, is illustrated in Figure 3.4. This identifies three potential sources of fear: novelty, innate threats and learned threats. Neophobia, or the fear of novelty, is expressed to a greater or lesser extent in all sentient animals, the extent being determined largely by their capacity to defend themselves from perceived threat. Success in life depends on acquiring the correct balance between curiosity (to develop survival skills) and fear (to avoid potential threats to health and safety). At birth, novelty is all since nothing is known. New-born mammals and birds begin life with little sense of fear but a great deal of curiosity. In nature, they are highly motivated to explore the environment under the watchful eye and protective cover of a parent in order to learn the skills necessary to promote welfare and increase the chances of survival. Among these skills is the ability to distinguish between threats that are real and imaginary. As the young animal builds up its store of experience in its natural habitat the chances of encountering new experiences become less and this reduces the 'need' for curiosity. At this stage neophobia, or at least a healthy caution in the presence of novelty, acquires a greater survival value than curiosity. In traditional and extensive farming systems, farm animals may not acquire survival skills in a natural way but they should become habituated to the normal sights and sounds of farm activity. They learn that the presence of the stockman or the sound of a tractor pose no real threat and may signal something good, like food. This habituation is an essential part of good husbandry. However, in some intensive systems, animals such as turkeys or veal calves reared in near darkness and given little or no opportunity to gain experience of life can panic at the slightest alarm. This arises from a failure to habituate and is hugely magnified by their innate recognition of the signs of fear in other conspecifics. Mob panic in intensive poultry houses is not just a welfare problem, it can result in the death of thousands of animals.

Figure 3.4 also illustrates the consequences of fear as a learning experience. An animal that experiences fear in response to a threat of any sort will, if it can, act to address the threat. Subsequently they will learn whether their action was effective, ineffective or unnecessary. In conditions of intensive confinement, opportunities for effective action (e.g. hiding or escape) may be restricted or non-existent, in which case the animals have no option but to wait and hope that the perceived threat will either go away or turn out not to be a problem. The animal that learns to cope with real threats or, through experience, overcome the problem of neophobia, will habituate, and the overall experience of fear will diminish. On the other hand, the animal that discovers that it is unable to cope, probably because the system denies it the possibility of normal behaviour (e.g. escape), then the acute, adaptive emotion of fear will escalate to a chronic non-adaptive state that may range from extreme anxiety to the profound depression of learned helplessness. I cite two extreme examples.

- Anxiety: Some laying hens in colony systems develop into 'pariahs'. These birds alternate between cowering in corners and rushing about in panic. Their condition deteriorates rapidly and then they die (or are culled).
- Learned helplessness: Working donkeys that are chronically beaten and ill-treated display many of the symptoms of learned helplessness, including the absence of any 'normal' response to painful stimuli.

The difference between acute fear and chronic anxiety (or learned helplessness) is a classic example of the difference between stress and suffering. Husbandry practices (e.g. handling, veterinary procedures) that cause acute fear are sometimes necessary. Practices that lead to chronic anxiety or learned helplessness are unnecessary and a cause of suffering. By this definition, they are unacceptable.

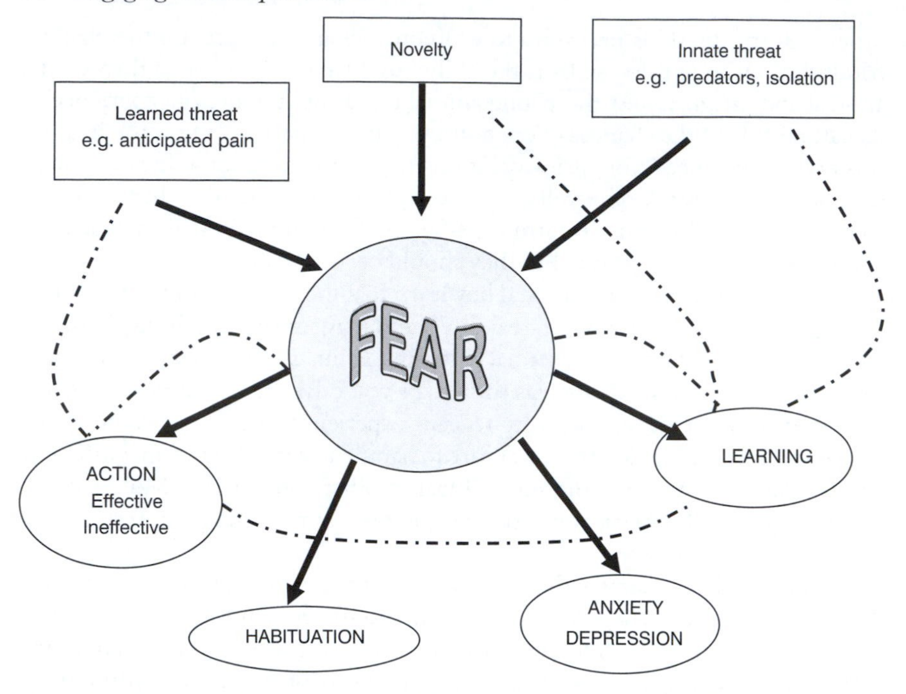

Figure 3.4 Causes and consequences of fear (from Webster 2005). The solid arrows indicate direct effects, the dashed lines indicate feedback loops

Frustration and deprivation

An emotional state of frustration can arise when animals are highly motivated to perform a specific behaviour but are unable to do so because of the presence of physical barriers or other constraints blocking access to key resources. Even the most highly selected commercial hens are highly motivated to seek a nest wherein to lay their eggs. Increased pacing is observed in hens that lack a nest site (but have room in which to pace). In the extreme confinement of the battery cage, they do not have the room to work out their frustration.

One of the indicators of frustration and deprivation is *redirected behaviour.* Important examples include feather pecking in hens and tail sucking/biting in pigs. In both cases the action is normal but directed towards the wrong object. It is normal for hens to use their beaks to investigate the environment. It is harmful when hens, surrounded by little more than other hens, constantly peck one another. It is normal for piglets to suck their mothers' teats and chew straw. It is harmful when piglets, weaned too early from their mothers and confined in barren cages suck and chew each other's tails.

The other important behavioural indicator of frustration and deprivation is *stereotypic behaviour.* This describes the persistent performance of invariant and apparent purposeless patterns of behaviour, such as pacing in zoo animals, bar-biting in confined sows, weaving and crib-biting in horses. Stereotypies such as pacing and weaving probably express a ritualized pointless expression of behaviour originally directed towards escape from confinement. Bar-biting and crib-biting are ritualized distortions of normal eating behaviour in animals motivated to forage for long hours but presented with meals that can be eaten within minutes.

It is a matter of dispute among scientists as to whether stereotypic behaviour is an expression of distress or a coping mechanism. We may not be sure whether these animals are suffering. We can be sure however that the expression of stereotypic behaviour is a clear indicator of an environment that fails to meet the animal's behavioural needs.

Health and happiness

Good welfare is more than simply the absence of disease, fear, pain and frustration. It should not need saying that sentient animals can experience feelings of contentment and pleasure. It is therefore proper that we should recognise behaviours associated with pleasure, such as play, mutual grooming and exploration, and seek, wherever possible, to create environments wherein these behaviours are observed. What these behaviours have in common is that they demand time and energy yet do not yield any immediate and obvious practical reward. In consequence, they tend to be the first activities to disappear when conditions get difficult. Clear evidence of these expressions of good health and happiness is a powerful confirmation of good husbandry.

Health and welfare problems in animal production systems

I conclude this chapter with brief lists of the most important problems of health and welfare that arise within intensive (Table 3.2) and extensive (Table 3.3) production systems and where the major risks are the production methods themselves. These lists could have been made much longer. They have been kept deliberately short in order to give emphasis to the top priorities for action to improve farm animal health and welfare. The problems marked by an asterisk are those where action is imperative. Most of these conditions are considered several times and in several places within this book. At this stage I shall discuss only some of the problems in order to illustrate general principles of the husbandry/health/welfare equation.

Intensive systems

Table 3.2 highlights the most important problems for pigs, poultry (broilers and laying hens) and dairy cattle (including veal calves) in intensive systems. The table links the problems to different aspects of husbandry: feeding, housing, breeding and management. In many cases the risks are multifactorial. Lameness in dairy cows, for example, is linked to deficiencies in feeding, housing, breeding and management. However the asterisk appears in the management

Table 3.2 Origins of major problems of health and welfare for pigs, poultry and dairy cattle in intensive production systems.

	Pigs	*Poultry*	*Dairy cattle (including calves)*
Feeding	Post weaning enteritis*	Lameness (Br)	Infertility* Ketosis Rumen acidosis Anaemia, Ulcers (veal)*
Housing	Enzootic pneumonia Stereotypies Aggression Tail biting Lameness	Lameness (Br) Frustration (LH) Bone fractures (LH) Feather pecking (LH)	Lameness Mastitis, Abnormal behaviour (veal)*
Breeding	Lameness (sows)	Lameness (Br)* Bone fractures (LH)* Aggression (LH) Feather pecking (LH)?	Infertility* Mastitis Lameness Exhaustion
Management	Aggression Pain following mutilations		Lameness* Pain following mutilations

* = most serious origins, Br = broilers, LH = laying hens

row because the main risks arise from failures in foot care and foot hygiene. In the case of infertility in dairy cows, both feeding and breeding get an asterisk, which reflects the fact (discussed in Chapter 2) that selection for milk yield has imposed a severe strain on the capacity of the cow to consume and digest enough feed to meet the metabolic demands of lactation.

The welfare problems of veal calves have been a subject of major concern and given rise to new EU legislation. Traditionally veal calves were confined in individual stalls and reared on an entirely liquid milk-replacer diet. This created a multitude of welfare problems of which the most severe were the constraints on natural behaviour and the abnormal diet, which induces anaemia, prevents normal rumen development, predisposes to severe indigestion, abomasal ulcers and enteric infections. European legislation now requires calves reared for veal to be reared in groups after the age of eight weeks and offered sufficient digestible fibre to promote rumen development. These are steps in the right direction but the life of the veal calf remains far from ideal.

Lameness, and the pain arising from lameness, present the main welfare problem for broiler chickens. In this case the lameness arises from abnormal skeletal development and secondary infections arising from abnormal strains on immature joints. Here too the risks are complex and can involve feeding systems, stocking density and litter management. Indisputably the greatest risk factor is the policy of the international breeding companies to select birds for rapid growth. They claim that in recent years they have given greater attention to skeletal development. However independent evidence (Knowles *et al.* 2008) reveals that the prevalence of lameness in broiler chickens has not significantly changed in the last fifteen years.

One of the greatest and most intractable welfare problems for laying hens is feather pecking. As indicated above, this is a form of redirected investigatory behaviour directed towards other chickens. There is good evidence that it is quite distinct from aggressive behaviour. Laying hens can be aggressive and some breeds are more aggressive than others. Aggression is usually directed at the head and neck and is accompanied by other indicators of conflict. The motivation for feather pecking remains obscure (to me it appears completely mindless), but it can do a lot of damage and, in extreme cases, lead to cannibalism when birds peck compulsively at exposed, inflamed and bleeding skin. The risk of feather pecking is greater for large numbers of birds in colony systems than for small numbers confined in cages. The preferred control method is beak trimming; removal of the tip of the upper beak. The details of this procedure and pros and cons of this practice are too complex to review here. However, the argument for beak trimming is that it is a mutilation carried out for the long-term benefit of the animals and, until we can breed birds that are not motivated to peck feathers, this argument stands.

Tail biting in pigs, like feather pecking in hens, is described as redirected behaviour, and a standard commercial approach to its control is tail docking, a practice designed to make the tail less conspicuous and thus less attractive to other pigs. However the justification for tail docking is far weaker than that for beak trimming since, in almost all cases, the problem can be overcome by more sympathetic approaches to management, especially the provision of an enriched environment that gives the pigs other interesting things to do with their mouths. Items such as footballs, rubber tyres and hanging chains were initially popular with farmers because they could be offered to pigs housed on concrete. However they were not popular with the pigs for the simple reason that they offered no lasting reward. The materials that attract and hold the attention of the pigs include straw, hay, wood, peat and mushroom compost in which they can engage in the natural behaviours of rooting and foraging. These things may be inconsistent with intensive systems that lack mechanisms for disposal of bedding. Yet it is beyond dispute that systems that deny pigs the expression of these natural patterns of behaviour seriously compromise both emotional and physical elements of their welfare.

Extensive, pastoral and village systems

Table 3.3 lists the most important problems of health and welfare that may occur in extensive systems. Here the word extensive covers a wide range of systems that include extensive pastoral systems for ruminants and small-scale 'village' systems of pig and poultry production both within the developing world and as practised by hobby farmers and the new class of middle-class peasants living in

Table 3.3 Origins of major problems of health and welfare for ruminants (cattle, sheep and goats), poultry and pigs in extensive systems, including organic and small, village-based systems

	Ruminants	*Poultry*	*Pigs*
Feeding	Hunger* Infertility		
Environment	Parasites* Infection Predation	Predation* Parasites* Infection Aggression, fear	Thermal stress
Management			Aggression
Breeding	Wrong phenotype	Wrong phenotype	Wrong phenotype

* = most serious origins

conditions of rural and suburban affluence. The problem list is shorter than that for intensive systems for the good reason (expounded already) that animals in extensive systems have more opportunity to make a constructive contribution to the quality of their own existence. However these animals can be faced by problems that are frequently severe and, in extreme cases, lethal.

The most common problem for grazing animals is hunger. Death from starvation in drought conditions, when feed runs out altogether, is clearly a disaster for all concerned. It does come within my purview when it is as a direct result of management policies. It is entirely natural for wild and domestic grazing animals to go hungry and lose body condition on a seasonal basis (dry season in the tropics, winter in the temperate and boreal zones). However, so long as the range or pasture is not overstocked, the animals should have some access to dormant grasses and shrubs, which may not provide enough nutrients to prevent loss of body condition but can permit the animals to make a constructive contribution to the quality of their own existence by foraging for long hours for some reward. It is valid to argue that the emotional state of a sheep or camel spending long hours scratching an inadequate supply of nutrients from dormant winter range is superior to that of a sow confined on concrete in a pregnancy stall and fed a quantity of food just sufficient to meet her nutrient needs, but consumed in a matter of minutes.

The most important environmental problems for free-living domestic animals are parasitism, predation and infection. All these problems are controllable to some degree through attention to management. However these things cost money. Vaccines and drugs for parasite control may be considered uneconomic by the hard-nosed commercial rancher. For the village farmer in the developing world their cost may be prohibitive. A comprehensive evaluation of the control of health and welfare in all the animals reared in all the different extensive systems in all the world is self-evidently way beyond the scope of this (or any other) book. The big message in the present context is that the most effective approach to the improvement of health and welfare in these systems is through the improved control of parasitism and infectious disease. In Chapter 1 I described the enormous boost to village chicken production in Africa through the supply and effective distribution of Newcastle disease vaccine. This is a beautiful example of small-scale, but well focused action leading to massive reward.

The last general issue that needs to be discussed at this stage is the question of breeding for the environment. The performance, health and welfare of any animal is detemined by how well its phenotype is suited to the environment to which it is exposed. As discussed in Chapter 2, in intensive systems, animals like the broiler chicken or the Holstein cow have been bred for maximum productivity within tightly controlled environments designed to minimise (especially) nutritional and thermal challenges to physical welfare. In extensive and environmentally challenging environments animal breeders have had to direct their breeding strategy to traits best suited to the environment in which they happen to live. Thus while the broiler chicken or the Holstein cow look much the same the world over, there are hundreds of breeds of sheep all different but mostly just right for where they live.

Problems have arisen when entrepreneurial or well-meaning individuals seek to introduce 'improved' animals into environments and systems to which they are clearly unsuited. A commercial, entrepreneurial, example of this involved the exportation of North American Holsteins, bred to eat starch and live in a barn, to the high-quality grasslands of New Zealand. An even less intelligent policy has been the well-meaning 'send a cow to Africa' scheme. It is beyond dispute that there is potential to improve the productivity of traditional breeds in the developing world through a well-planned breeding policy that achieves the genetic mix most suited to achieve optimal productivity without compromise to health and welfare within specific environments carefully described in terms of all the physical, social and biological parameters ilustrated in Figures 3.2 and 3.3. In Chapter 6 I shall present in some detail two approaches to improving poultry production and dairy production in developing countries. I shall close this chapter by requoting Einstein: 'Everything should be as simple as possible but no simpler'.

Notes

1 Nociceptive is a term used by physiologists to describe reflex or conscious evidence of response to a painful stimulus but not its complex consequences for a sentient animal.

2 Statistics from the National CJD Research and Surveillance Unit (NCJDRSU), Edinburgh, UK.

4 Food from animals

> Give a man enough food and he has many problems, give a man not enough food and he has only one problem.
>
> (Proverb)

The proverb that prefaces this chapter neatly encapsulates the fact that our view of food depends on where we stand. For those living at or below the subsistence level, the quest for food, any sort of food, is, most of the time, the great motivator (the one problem). Success brings the reward of a good meal, or at least temporary relief from the pangs of hunger. Failure leads inevitably to physical decline, accompanied by the behavioural signs of failure to cope. With chronic hunger as with fear (Figure 3.4) these signs may range over a wide spectrum from pathological anxiety to extreme apathy and learned helplessness.

For those who have year-round access to a wide variety of attractive food, and can afford to buy it, food is one of the great pleasures of life. We love food, the taste, smell and appearance of food, the art of preparing food and the social satisfaction to be derived from sharing a good meal. George Bernard Shaw wrote that 'the love of food is the most sincere form of love' but it is a love that carries a heavy burden of fear and guilt. We fear the diseases and disorders linked to the sin of gluttony, such as cardiovascular disease (CVD), cancer, diabetes and obesity. We feel (or should feel) guilt in matters relating to the exploitation of animals, farmers in the third world and the natural environment. In short, we have many problems. These sources of fear and guilt are valid but our approach to these problems tends to be driven more by emotion than reason (sentience dominating cognition). The aim of this chapter is to explore our relationship with food from animals in a way that is rational and based on sound science, but recognises (as a scientist) the power of emotion as a motivator of human behaviour, and recognises (as one who is trying to get results) the importance of working in sympathy with humans as we are. As in all these chapters, I shall not attempt a brief (and therefore superficial) overview of all the associations between food from animals and human nutrition, health and behaviour. Instead I shall concentrate on the big issues in the context of current practices in animal husbandry and opportunities to do things better.

The value of food from animals: direct measures

The first step is to set out the factors that determine how we value our food and who sets these values. For many humans and most free-living animals, the need to gather enough food for survival is the big problem so all food is likely to be of great value. However, when there is some choice, both animals and humans have the instinctive physiological ability to select a diet that matches their needs for specific nutrients. Since most food is used as fuel for energy metabolism, one of the main drivers of food selection is energy density, which explains our instinctive attraction to foods high in fats and oils (Bludell *et al.* 1996). However we are also instinctively driven to select foods that meet requirements for specific nutrients. This has been studied more intensively in animals than humans. Given free choice, growing rats and pigs will select diets and adjust their appetites to provide sufficient protein (essential amino acids) for optimal growth of functional lean body mass. If the diet is deficient in protein concentration they will eat more in the attempt to meet protein requirement for optimal lean body growth and, in the process, deposit excessive amounts of body fat (Webster 1986). This instinctive need to meet protein requirement is a significant contributor to obesity, whether in affluent Western children with easy access to junk food, or African women with easy access to starch but limited access to protein. There is a fascinating literature on diet selection to meet nutrient needs in animals – sheep selecting higher protein grasses to offset losses arising from intestinal parasites, range cattle seeking to remedy phosphorus deficiency by chewing bones – but this is outside the scope of this argument. The message here is that, when food is hard to get, its value is primarily determined by its ability to keep us alive and well in the short term; i.e. to meet our immediate needs for nutrients to support the essential functions of maintenance, growth, work, reproduction and lactation. In these circumstances, when foods of animal origin, meat, milk and eggs, are scarce or prohibitively expensive, their real value is their nutritive value, their capacity to reinforce diets based on starchy cereals and tubers with essential amino acids, fatty acids, minerals and vitamins.

In the affluent world, as distinct from the poor world (I try to avoid the patronising use of 'developed' and 'developing'), things are not so simple. We measure the value of food in many different ways (Box 4.1). The first set of values is based on direct assessment: 'Is this food worth buying, whatever its provenance?' The people who set these values are, by definition, the customers and consumers. Each of us who buys food makes a series of more or less informed decisions in regard to appearance, taste, cost and convenience. I put appearance before taste on the basis of overwhelming evidence that this is given greater importance when selecting food from the supermarket shelf, if not at the time of the meal. Cost is obviously a major factor in food selection, whether our estimate of value be direct (stew vs. steak) or indirect (battery eggs vs. free-range). So too is convenience. One of the undisputed attractions of supermarkets is the fact that all the food shopping can be done at one fell swoop.

Box 4.1 Measures of food value

Direct measures

Consumer-based

- Appearance, taste, cost, convenience
- Food safety
- Perceived health risks and benefits.

Science-based

- Nutrient supply: meat, milk and eggs as components of a mixed diet
- Health risks: obesity, diabetes, cardiovascular diseases and cancers
- Statistical associations: obesity, fatty acids, cholesterol and CHD, meat and cancers
- Functional associations: integrity of the digestive tract, cellular integrity.

Indirect measures

- Provenance: local produce, organic standards, fair trade
- Production standards: sustainability, organic, animal welfare, farmer welfare
- Advertising and social pressure.

The next listed criterion of food value is food safety. Clearly it is the responsibility of all involved in the food chain from farm to fork to minimise the risk of contaminating food with potentially harmful microorganisms, toxic chemicals or antibiotic residues. However, customers have to take these things on trust. Retailers, especially those selling fresh and processed meats that are often the source of outbreaks of food poisoning, have a legal obligation to operate to the highest possible standards of food hygiene. They are also aware that all fresh food, particularly meat and fish, needs to be presented in a way that looks hygienic. Far more people buy their meat in vacuum packs from the supermarket than have it chopped up for them by the butcher. Reasons for their choice include, of course, cost and convenience, but it is fair to assume that they are also motivated, consciously or instinctively, by the belief that such food looks 'safer' or less obviously part of a once-living animal.

The third direct measure of value is based on consumer-based perception of longer term risks and benefits to health associated with the consumption of specific foods. Consumers are becoming increasingly aware of the hazards of unwise eating, such as obesity, diabetes, cancer and coronary heart disease (CHD). At the same time we are bombarded with stories of foods that are good for us and foods that are bad for us. This is a confusing and potentially dangerous mixture of good and bad science, good, bad and dreadful journalism, and outright quackery, promoted by individuals ranging from the cynical seller of supplements to the simply deluded. (These can be the same people, since if the aim is to fool all of the people all of the time, it helps to fool oneself first.)

Consumer confusion regarding the associations between food and health is a consequence of too little education and too much propaganda. However this in itself reflects a lack of understanding and agreement among nutrition scientists. While we have a thorough knowledge of nutrient supply and nutrient requirement, we know remarkably little of what we need to know about the long-term associations between nutrition and health. About the only thing we can say with absolute confidence is that it is almost meaningless to talk of good and bad foods. They can only be considered within the context of the overall diet.

The science of human nutrition in the context of human health proceeds along two parallel fronts. The first approach is epidemiological, and looks for statistical associations between elements of diet as input (risk) factors (see Figure 3.2) and measures of mortality and morbidity (e.g. cancers, CHD, diabetes) as outcome measures. The advantage of this approach is that one can study large populations. The limitations, uncertainties and potential for error are legion. Statistical associations are not measures of cause and effect. Apparent associations between hypothesised cause (e.g. saturated fats) and predicted effect (CHD) will always be confounded by other nutritional, genetic, environmental and random factors. These can be reduced but not eliminated by good statistical design. Epidemiological studies with human populations are also plagued by uncertainty as to what and how much the people are actually eating (even when they write down what they allege they have eaten). Finally, there is the problem of scientific bias. Scientists are under constant pressure to go to press, especially when they appear to have revealed a statistically significant difference between diet A (containing substance X) and B. On the other hand, when their experiment fails to reveal a difference between diets A and B, then they, and their sponsors, are less keen to publish. This inevitably creates a bias: too many small studies that suggest, but do not prove, that Substance X is good, or bad for you, or both; and too few to suggest that the consumption of Substance X in sensible quantities is likely to have no significant consequences for health, one way or the other. This bad science is cynically exploited by journalists with conveniently short memories and the editorial brief to 'chill their readers to the very marrow'.

The relatively new statistical science of meta-analysis can do much to address this serious problem. Meta-analysis is a statistical technique for combining the results from independent randomised control trials, giving due weight to the scale of the different studies. At best such studies can give a more robust estimate of treatment effects and account for differences between individual studies. However they too are subject to bias. They may, for example, omit (for whatever reason) individual well-designed studies. More commonly, in these days of transparency and online access to all the literature, they may be obliged (e.g. by honestly misguided governmental sponsors) to take into account *all* the literature, including papers that a competent scientist should be allowed to discard as clearly inadequate, usually on the grounds of poor design.

Epidemiological science is necessary to highlight potential problems and can sometimes tell us all we need to know. The statistical association between smoking and lung cancer is now universally acknowledged as cause and effect. It is confirmed by evidence that tobacco tars have carcinogenic effects in experimental animals that are proportional to the dose (i.e. all tobacco is bad and the more you smoke the worse it is). With foods and food substances, things are less straightforward. An essential nutrient (e.g. vitamin and/or provitamin A) must be eaten in small quantities. Having met this requirement, further intake of vitamin A in the form of supplements is likely to be harmless but unnecessary, up to a point. Excessive intake of vitamin A becomes toxic and, in extreme cases, lethal. (The Inuit are well aware of the dangers of eating the vitamin A-rich livers of seals and especially polar bears.) Thus it is meaningless to speak of (or attempt to measure) the benefits of vitamin supplements without good evidence of their availability in the basal diet.

The main thrust of current human nutrition science is to discover, by experiment, functional and mechanistic reasons why certain food ingredients may do good or harm. Here again, the number and range of studies is enormous but most can be placed within one of two categories:

- effects on the microenvironment and immunological status of the digestive tract;
- effects on the functional integrity of cells (including the intestinal epithelium).

I shall consider these in more detail when examining potential risks and benefits associated with eating meat, milk and eggs.

The value of food from animals: indirect measures

Box 4.1 lists three indirect measures of food value open to those who have the luxury of choice: provenance, production standards and social pressure. The third of these is undoubtedly the strongest. Advertising brings certain foods, and makers of these foods, to our attention. Advertising may put emphasis on the health-giving properties of the food 'when eaten as part of a balanced diet'. In Europe these claims are subject to quite strict legislation. More commonly the advertising will suggest that your choice of this particular food or drink, or this particular supermarket, implies somehow that you are a special, superior or sexier person. This can be the only explanation for a class of citizen that is prepared to pay more for bottled water than for milk.

Provenance is potentially a more valid indirect measure of food value, so long as the claims can be substantiated by independent audit. The appeal of local produce is based on sound foundations: freshness, low food miles, support for local farmers ('our people') and, in the case of food from animals, trust that 'our people' will rear their animals in a way that commands our respect. 'Fair trade' is another increasingly popular measure of provenance, the principal aim being to ensure fair treatment for producers and sustainable management of the environment, particularly in the third world. To date however, products marketed under the fair trade label are mostly luxury or semi-luxury items such as coffee, chocolate, bananas, rather than essential, staple foods. To my knowledge, there has been no significant movement to market foods of animal origin under the fair trade label.

Organic foods are listed in Box 4.1, under both provenance and production standards. Today, organic food production has become a highly regulated industry for the production of food using methods that 'do not involve modern synthetic inputs such as pesticides and chemical fertilizers, do not contain genetically modified organisms, and are not processed using irradiation, industrial solvents, or chemical food additives' (Soil Association n.d.). It should be fairly clear that these guarantees are primarily directed at perceived fears as to risks to human health from eating food that has been produced by 'unnatural' methods. This is consistent with the original aims of the mother of organic production, namely the UK Soil Association. Sir Albert Howard, one of its principal sources of inspiration, wrote in 1940 that 'only healthy soil, produces the right food to help man's health'. The overall weight of evidence, based on meta-analysis, indicates that conventional foods and foods of organic origin do not differ significantly either in nutritive value or in effects on health (Bourn and Prescott 2002).

Despite the weakness of the original premise, there is much to commend in the philosophy of organic production. Indeed, the farming methods they advocate are almost entirely consistent with the central theme of this book, namely the efficient, sympathetic, sustainable husbandry of farmland to the lasting benefit of all. Animal husbandry is essential to organic farming, most obviously as a source of organic fertilizer, but also because mixed farming is frequently the most biologically efficient form of land use. Land that is best suited to the production of grass, or agroforestry, should be grazed by herbivores, and we should manage these herbivores in such a way as to produce a yield of meat, milk, leather, etc., while sustaining the health of the living environment (animals, plants, soil and water).

The other big indirect measure of food value according to production standards is, of course, animal welfare. In Europe and North America there is growing consumer demand for food that is quality assured as 'High Welfare' and a number of regulatory bodies have been set up to police and promote these standards (e.g. RSPCA 'Freedom Foods' in the UK). Surveys of food-buying habits, taken during the 2009–10 recession indicate that sales of high-welfare food are proving more robust than sales of organic food. There may be several reasons for this, not least that food sold under the 'high-welfare' label tends to be cheaper than the full organic package (which includes high-welfare). Nevertheless, the evidence does suggest that 'high-welfare' can stand alone as a measure of added value in food from animals. Discussion of the assessment, surveillance and promotion of farm animal welfare appears in almost every chapter of this book. The point of emphasis here is that welfare has been recognised as an increasingly important measure of the value of food from animals, therefore an increasingly important incentive to good husbandry.

The composition of food from animals

Table 4.1 describes the main constituents of foods from animals, beef, chicken, milk, cheese and eggs, together with a staple cereal, namely rice. The values are taken from the comprehensive data sets provided by the USDA Food and Nutrition Information Centre (n.d.). These reveal considerable variation within a product (e.g. beef) according to location on the carcass, fat trim, methods of preservation and cooking. In Table 4.1 all columns describe raw material 'straight from the animal'. The beef example describes a sirloin steak with 18mm fat trim. The chicken meat example describes boned chicken breast. The first row lists the water concentration (as a percentage) of each product and shows that the water content of lean meat (chicken, trout) and eggs is similar at about 75 per cent. The lower water concentration of beef steak in this example reflects the greater amount of fat retained after trimming. The subsequent rows, unconventionally, compare nutrient concentrations per 100g dry matter (DM) or water-free mass. This is to avoid confounding comparisons between products in terms of constituents such as protein, saturated fats and cholesterol in terms of nutritive value or possible health risks with differences resulting simply from differences in water content which are not relevant to the present argument.

All the listed animal products (meat, fish, milk, cheese, eggs) are broadly similar when measured as sources of food energy (i.e. 2,000–2,500 kJ/100g DM). The proportion of food energy derived from fat (at 39kJ/g) ranges from 21 per cent for lean chicken breast to 51 per cent for beef steak and 76 per cent for Cheddar cheese. Meat, cheese and eggs contain negligible amounts of carbohydrate. The concentration of protein in the listed animal products ranges from 28 per cent (eggs) to 88 per cent (lean chicken breast). In all cases this is far in excess of the concentration of protein required for optimal human nutrition (10–18 per cent, see Chapter 1). Only a small amount of meat, dried milk, cheese or eggs is necessary to meet human protein requirements (even those of growing children or lactating mothers) subsisting on a diet of rice (8 per cent protein in DM). Most well-fed omnivores eat more food from animals than we need to meet both protein and energy requirements. The big biological and sociological issues revolve around the question 'how much is too much?'

Much of the concern regarding food from animals and human health relates to the quantity and composition of animal fats (lipids) which are essentially complex molecules made up of chains of fatty acids bound together by glycerol. Most of the animal fat we eat acts as a high-density source of fuel for energy metabolism, to be used at once, or more likely, transferred to our own fat reserves to meet future needs. Regular intake of ME, principally from fats and sugars, in excess of energy expenditure leads to obesity. In recent years however there has been an explosion of interest in the fatty acid composition of foods:

Table 4.1 Composition of meats, dairy products, eggs and rice (from USDA Food and Nutrition Information Center). The first row lists water concentration in the raw state (%). All other values are presented as proportion of the water-free mass (dry matter, DM)

	Beef, steak	*Chicken breast*	*Trout, whole*	*Milk, Holstein*	*Cheese, Cheddar*	*Eggs*	*Rice*
Water (%)	68	76	74	88	37	76	12
Composition/100g DM							
Energy kJ	2466	1990	2265	2233	2673	2496	1735
Energy kcal	591	475	542	533	640	596	415
Protein	65	88.3	76.9	27.3	39.5	52.5	8.1
Carbohydrate (g)	0	0	0	38.3	2.0	3.0	91
Total fat (g)	35.0	10.8	23.8	30.5	52.5	39.6	0.75
Saturates	14.0	2.4	5.4	18.6	33.5	13	0.2
Monounsaturates	15	3.2	7.7	9.7	14.9	15.25	0.24
Polyunsaturates	1.3	1.7	5.8	1.2	1.49	7.96	0.2
Cholesterol (mg)	225	267	226	117	167	1550	0
Calcium (mg)	75	21	96	992	1144	233	32
Potassium	606	1542	1450	1258	156	575	131
Sodium	166	483	196	408	986	592	5.7
Iron	4.7	1.96	1.2	0.42	1.08	7.29	0.91

which fatty acids are essential to life, and how different classes of fatty acid may impact on human health. Table 4.1 presents a broad classification of these fatty acids as saturated, monounsaturated and polyunsaturated. Meat and dairy products from ruminants contain a relatively higher proportion of saturated fats than meat from simple-stomached animals such as pork and poultry. This reflects the fact that the flow of digestible nutrients to ruminants is dominated by the anaerobic process of fermentation in the rumen, which tends to saturate unsaturated fatty acids present in the diet. The composition of fats in simple-stomached animals like pigs and poultry is much more dependent on the composition of the diet, so open to manipulation (see below). All the animal products listed in Table 4.1 contain significant amounts of monounsaturated fatty acids. Meat from mammals and birds (beef, lamb, pork, poultry) and dairy products contain only small amounts of polyunsaturated fatty acids (PUFA). However eggs and fish (especially oily fish like trout) are PUFA-rich.

Box 4.2 presents a fuller classification of the fatty acids and their dietary sources. As indicated already, dairy products and meat, especially from ruminants, are overwhelmingly the main sources of saturated fats in the diet of the human omnivore. The proportion of monounsaturated fatty acids in food from animals is within the range of that found in vegetable oils, with the exception of olive oil which is >90 per cent monounsaturated. The polyunsaturated fatty acids, which include the essential fatty acids, are subdivided into the ω-3 (or n-3) and ω-6 (or n-6) classes according to the position of the first, unsaturated double bond (i.e. CH=CH, as distinct from saturated CH_2-CH_2 bonds). The ratio of ω-3 to ω-6 PUFA in fish oils is about 0.6:0.4. The fish do not manufacture these ω-3 PUFA themselves but obtain them from plankton and marine algae. In conventionally produced hens' eggs the ratio is about 0.4:0.6. The implications of ω-3 to ω-6 ratios for human health, and the manipulation of these ratios in animal products (principally eggs) are discussed below.

The *essential fatty acids (EFA)* are a subset of PUFA that cannot be synthesised by humans, as humans lack the desaturase enzymes required for their production. They include the short-chain polyunsaturated fatty acids (SC-PUFA) a-linolenic acid (ALA, ω-3) and linoleic acid (LA, ω-6). These two fatty acids form the starting point for the creation of long-chain PUFA, especially the *eicosanoids* which are directly involved in responses to cell damage, i.e. pain, inflammation and repair. In general the eicosanoids synthesised from ω-6 fatty acids are more inflammatory, which may help in combating damage accompanied by infection. Eicosanoids synthesised from the ω-3 fatty acids are associated with less extreme inflammatory reactions that are more appropriate when the damage is chronic and self-inflicted (e.g. atherosclerosis as a precursor of CVD). Consumption of high quantities of the ω-3 essential fatty acids from fish and marine mammals (seals) is a convincing explanation for the 'Eskimo (Inuit) paradox'; namely the low incidence of CVD in a population that consumes prodigious quantities of fat. However, given our current obsession with ω-3 PUFA, it is essential to restate the fact that both ω-3 and ω-6 EFA are, by definition, essential.

Box 4.2 Fatty acids: composition and sources

Saturated fatty acids

Examples: Palmitic acid $CH_3(CH_2)_{14}COOH$
Stearic acid $CH_3(CH_2)_{16}COOH$
Sources: Meat and dairy products, some vegetable oils (e.g. palm oil).

Monounsaturated fatty acids

Example: Oleic acid $CH_3(CH_2)_7CH{=}CH(CH_2)_7\ COOH$
Sources: Olive oil >90%, other vegetable oils 20–60%, animal fats 40–55%.

Polyunsaturated fatty acids

Omega-3

Examples: a-linolenic acid (ALA, *all-cis*-9,12,15)
Eicosapentaenoic acid (EPA, *all-cis*-5,8,11,14,17)
Sources: ALA: oil seeds (soya, rape, linseed), EPA: fish oils.

Omega-6

Examples: linoleic acid (LA *all-cis*-9,12)
g-linoleic (GLA *all-cis*-6,9,12)
Sources: LA: Oilseeds (soya 50 per cent, sunflower 70 per cent), GLA, evening primrose.

Essential fatty acids

Trans fatty acids

Conjugated fatty acids

Example: Conjugated linoleic (9Z,11E-octadeca-9,11-dienoic) acid
Sources: meat and milk from (especially) grass-fed ruminants.

Trans fats are PUFA with one or more of their double bonds in the 'trans' rather than the common 'cis' configuration. They occur naturally at low levels in dairy products and meat from ruminant animals. They are also produced in higher concentrations during the industrial hydrogenation of vegetable oils (margarines, biscuits, pastries). Associations have been drawn between the consumption of trans fats and cancers of the colon and breast. At present these are tenuous, but there is already strong pressure to reduce or eliminate trans fats from vegetable oil products. There is no convincing evidence to link the low concentration of trans fats in animal products to cancer in humans (Thompson *et al*. 2008).

Conjugated fatty acids, especially conjugated linoleic acid (CLA) are present in milk and carcass fat from ruminants, with higher concentrations in animals fed from pasture than those fed diets containing a high proportion of cereals. CLA has been shown experimentally to reduce atherosclerosis and low-density lipoproteins (LDL) in rabbits. It inhibits lipogenesis and increases lipolysis in mice. The secondary effects of this are to reduce total body fat and appetite. As yet there is insufficient evidence to link CLA intake with appetite, body fat and the incidence of CVD in humans, but the experimental evidence is promising.

Food from animals and human health

There are major concerns, more or less substantiated by reputable scientific evidence, that the consumption of foods of animal origin increases the risk of acquiring diseases and disorders such as obesity, type-2 diabetes, cardiovascular diseases and a range of cancers (WHO 2003). I shall consider each of these in order.

Obesity

Obesity is the major food-related disease of those who can afford to eat more than they should. The explanation of cause and effect is blindingly obvious and should not be dodged. Food energy (ME) intake in excess of energy expenditure (H) and energy deposition in vital tissues such as milk and lean body mass will be deposited as fat, and the more and longer ME intake exceeds energy requirement, the fatter we shall get. Indisputably there are physiological differences between individuals in the predisposition to deposit fat relative to protein, and these interact with appetite (humans and strains of experimental animal predisposed to lay down excess amounts of fat are hungrier). Nevertheless, nobody is exempt from the consequences of eating too much. In this regard, foods of animal origin are high risk both because they are energy-rich and because they are very tasty. There is the counter-argument that high-protein feeds (e.g. lean meat) can keep you slim. This has been the basis for several intermittently fashionable diets, e.g. the Atkins Diet (Gardner *et al.* 2007) and is reasonably soundly based on experimental evidence of the satiating effects of protein in experimental animals. However in view of the other potential risks associated with high meat consumption (see below), the healthier solution must be to reduce overall ME intake and increase energy expenditure.

Obesity, typically caused by high intake of sugars and fats, is a major risk factor for the onset of type-2 diabetes. Overeating high-fat foods of animal origin obviously increases the risk of diabetes. However there is no convincing evidence of differences in risk attributable to fat class (i.e. saturated, mono-, polyunsaturated, ω-3 vs. ω-6).

Cardiovascular disease

The evidence for links between diet, obesity and cardiovascular diseases (CVD: atherosclerosis, stroke and CHD) is very strong. The classic early multinational studies of Keys (1980) identified a highly significant association between CVD incidence and blood cholesterol concentration, which was itself positively associated with dietary intake of dietary cholesterol and saturated fat and negatively associated with intake of unsaturated fat. Since a substantial amount of the fat in meat and dairy products is saturated (whereas fats and oils in plants are almost entirely unsaturated) and only animal-based products contain dietary cholesterol, consumption of meat and dairy products – particularly high-fat meat and dairy products – has come to be linked with an increased risk of CVD. Metabolic studies have subsequently established that the type of fat, but not total amount of fat, predicts serum cholesterol levels. In addition, results from epidemiologic studies and controlled clinical trials have indicated that replacing saturated fat with unsaturated fat is more effective in lowering risk of CHD than simply reducing total fat consumption. Prospective cohort studies and secondary prevention trials have provided strong evidence that increasing intake of ω-3 fatty acids from fish or plant sources substantially lowers risk of cardiovascular mortality (Hu *et al*. 2001). The correlation between dietary intake of cholesterol and blood cholesterol concentration is low, except in certain individuals with a genetic predisposition to high blood cholesterol.

The main practical conclusions to be drawn from this evidence are as follows.

- The main risk factor is the proportion, rather than the total quantity of saturated fat in the diet. Thus the risks from red meat and poultry meat are the same.
- Consumption of ω-3 PUFA has a protective effect against CVD (the Eskimo effect).
- Consumption of dietary cholesterol is not linked to blood cholesterol in most people; thus eggs are not a high-risk food.

Another risk factor for CVD and high blood pressure is salt intake (NaCl). Most balanced diets provide adequate sodium through its natural occurrence in meat, fish, eggs, fruits and vegetables. However, salt is used as a preservative and flavour enhancer in most processed foods (including cheese, bacon, ham, sausages, ready meals). The consumption of these processed foods is high and increasing. It has been calculated that reducing average salt intakes in the UK from current levels at around 9g per day to the recommended 6g per day would save about 8,000 deaths a year (Scientific Advisory Committee on Nutritiom 2003).

Cancer

The World Cancer Research Fund (WCRF 2007) has produced a comprehensive review of the epidemiological and mechanistic evidence to link increased and decreased risk of cancer to food, nutrition and physical activity. They identify three degrees of association: *convincing, probable* and *limited/suggestive.* Despite the massive amount of evidence on review (approximately 5,000 references) few of the associations are characterised as convincing or probable. The most convincing and consistent links relate to colorectal cancer. Factors identified as convincing contributors to increased risk include obesity (especially abdominal obesity), consumption of alcoholic drinks and consumption of red and processed meats. On the plus side, milk consumption probably reduces the risk of colorectal cancer. Obesity is also identified as a convincing or probable risk factor for cancers of the breast (in post-menopausal women), pancreas, oesophagus and kidney. There is no convincing evidence that fats *per se* directly affect the risk of cancer one way or the other, although fat consumption is, of course, a predisposing factor for obesity.

The mechanistic links between colorectal cancer and the consumption of red meats (beef, lamb and pork) are not fully understood. However there is good evidence that iron-containing haem products (e.g. myoglobin) that give red meats their colour are converted to known carcinogens (especially *N*-nitroso compounds) during cooking, especially high-temperature cooking. Moreover, burnt fats can also become carcinogenic. This makes barbecued steak the riskiest of all meats.

Meat processing involves salting, curing, smoking and treatment with preservatives including nitrites, all intended to prevent contamination of the meat with food spoilage bacteria. All these processes can generate *N*-nitroso compounds, which are known carcinogens.

The World Cancer Research Fund (WCRF) concludes that there is no convincing evidence for direct links between cancers and the consumption of poultry meat or eggs.

Manipulating the composition of food from animals

It is necessary, at this gloomy stage, to remind the reader that meat, milk and eggs are rich in essential nutrients. When eaten in relatively small amounts, they can provide all the essential ingredients necessary to supplement staple diets like rice and tubers which are rich in energy but not much else. Our need for these nutrients increases in proportion to our requirements for tissue synthesis (in addition to fuel for energy metabolism). The two neediest groups are growing children and lactating mothers. There is a cast-iron case for inclusion of small quantities of foods of animal origin, efficiently and humanely produced, in the diets of the poor and hungry. However it is equally beyond dispute that most of us who can afford it eat more meat, milk and eggs than we need, and in many cases, more than we should. Recognising this, the livestock industry, and scientists working in support of the livestock industry, are constantly investigating ways of making food from animals healthier, or perceived to be healthier, which, for marketing purposes, is much the same thing.

Meat and fish

Fresh meat is muscle, consisting principally of water (60–70 per cent), protein (20–5 per cent) and fat (5–15 per cent). Processed meats, sausages and other foods may include blood, offal, and preservatives (e.g. nitrites) but, in general, the aim is to produce a balance of protein, fat and water similar to that of meat since that is what we have become accustomed to like. Moreover, I repeat, we like the taste of fat. Even those who reject the strip of solid fat on the outside of a steak or a lamb chop, appreciate, consciously or unconsciously, the taste of 'marbling' fat within the muscle. Very lean meat, e.g. venison from wild deer, is generally perceived as 'dry' and calls for special skills in cooking. Animal breeders, especially traditional cattle breeders producing beef off grass, selected their animals not only for ease of fattening but also for fat distribution within the carcass. The Aberdeen Angus breeders, in particular, pride themselves on the ability of the breed to deposit intramuscular ('marbling') fat.

It is relatively straightforward to manipulate the total amount of fat in the carcass of ruminants through nutrition and breed selection. It is more difficult to reduce the proportion of saturated fats because, as indicated earlier, fat composition in these animals is dominated by the substrates resulting from anaerobic digestion in the rumen. There is some evidence that the proportion of saturated fats in beef is lower in animals finished on high-cereal rations than at pasture. On the other hand, there is evidence that finishing beef cattle on grass (which contains a significant amount of ALA) will increase the proportion of ω-3 PUFA and CLA in carcass fat. It must be stressed, however, that these differences are small and there is as yet no evidence that they can alter risks to human health to a degree that the report of the World Cancer Research Fund would class even as suggestive. If you want to significantly increase ω-3 PUFA intake, eat oily fish.

It is easier to manipulate the fatty acid composition of simple-stomached animals (e.g. pigs, poultry and fish). The ω-3 fatty acid content of both poultry meat and eggs can be readily increased by the dietary inclusion of marine oils/meals, which contain the long-chain ω-3 fatty acids (eicosapentaenoic acid 20:5n3, decosahexaenoic acid 22:6n3) found in marine sources. However, this practice has given rise to some problems arising from off-flavours in meat and eggs (and public concern relating to the feeding of fish to poultry). Feeding of short-chain ALA from vegetable sources has been less effective, producing only minor changes in the content of 20-carbon ω-3 fatty acids.

Eggs

As stated above, it is possible to elevate the ω-3 fatty acid concentration of eggs through the incorporation of fish oils in diets for laying hens. This has increased the demand for these eggs from a section of health-conscious consumers (the 'anxious well'). Whether it will have long-term consequences for their health may remain uncertain for years, not least because this section of society are also likely to be cautious in their consumption of high-risk foods, such as saturated fats, and avid consumers of a wide range of perceived health foods. There is no convincing evidence of differences in the composition of eggs from caged and free-range hens fed similar diets.

Milk

There is great potential to manipulate the composition of milk products, mostly after the milk leaves the animal. Butter, cream, ice cream, yoghourt and, especially, an enormous range of cheeses present splendid opportunities to increase demand and add value. Even low-fat skim milk represents added value (to the retailers) since they charge the same price for less nutrients and more water than in whole milk. To these we may add products such as spreadable butter, in which milk solids are supplemented with vegetable oils. Given the opportunities to manipulate milk products after leaving the animal in ways that increase the range of desirable dairy products and may have positive effects on health, it may seem relatively pointless to use dietary manipulation (e.g. the feeding of unsaturated fats in a form that bypasses the hydrogenating bacteria in the rumen) in order to manipulate the composition of milk as it leaves the cow. That has not inhibited the actions of entrepreneurs seeking market opportunities, e.g. the promotion of milk from cows milked in the morning, containing marginally higher levels of melatonin, to insomniacs seeking more satisfying bedtime drinks.

There are however some differences in milk composition between species and between breeds within species that may have implications for human health. Probably the most interesting of these are differences in the structure of the milk protein, casein. There are, in fact, several milk caseins, the principal being beta casein, which exists in two forms, A1 and A2 beta casein. Most black and white dairy cows (Holsteins and Friesians) secrete A1 and A2 in approximately equal proportions. Other species (e.g. goats, camels) and some breeds of dairy cow (conspicuously the Guernsey breed) secrete approximately 80 per cent A2 beta casein. A number of papers have emerged to suggest that milk containing substantial amounts of A1 beta casein carries an increased risk for CVD and type-2 diabetes (Kaminski *et al.* 2007). There is not yet sufficient evidence to permit effective meta-analysis or justify these health claims. Unsurprisingly, this has not inhibited the promotion of goat milk, and milk from Guernsey cows, on health grounds. Time may indeed prove them right. If so, it is feasible and may be sensible to select black and white cows with higher A2:A1 ratios. It is important to stress that differences in A2:A1 ratios have no relevance to the allergenicity of milk to certain hypersensitive individuals.

Conclusions: food production in the context of planet husbandry

Consumption of small quantities of meat, eggs and dairy products can make vital contributions to meeting human nutrient requirements, especially those for whom all food is in short supply and where the food that is available, cereals and tubers, is likely to consist of little more than starch. However, most of us with the luxuries of adequate income and freedom of choice consume more food of animal origin than we need. Consumed in moderate excess of need, and accompanied by reduced consumption of other energy-rich foods (e.g. cakes and biscuits) this will not necessarily lead to obesity or the diseases of affluence (e.g. CVD and cancers). Consumption of animal products (especially red meat and saturated fat) in substantial excess of requirement does significantly increase the risks of (especially) colorectal cancer and CVD.

There is real potential to reduce the risks of cancer and CVD through diet selection within foods of animal origin (e.g. trout rather than beef). There may be some potential to reduce these risks through manipulation of animal feeds (e.g. 'high ω-3' eggs, high CLA milk from cows at grass) or genotype (high A2 beta casein in Guernsey cows). Yet the simplest solution is to eat less meat and dairy products. This is, of course, entirely consistent with world needs to reduce animal production overall, and to reduce the feeding of animals on 'competitive' feeds (cereals and soya) relative to 'complementary' feeds such as pasture, forage crops and byproducts (Chapter 2).

An inevitable consequence of reduced output of food from animals per hectare or per farm will be reduced income from the sale of food and this is a loss that farmers cannot be expected to bear in current economic circumstances. This can be overcome in part through selling 'health' foods, or sustainably produced foods at higher cost, but it is unrealistic to expect that this alone will support the reduction in output needed by each of us as individuals, and the world at large. The future for good animal husbandry must depend on integrated forms of land use that combine food production with the production and maintenance of other elements of value such as carbon conservation, water and wildlife management, amenity and, of course, beauty.

5 Nature's social union

Philosophy, politics and economics

> The great fault of all ethics hitherto has been that they believed themselves to have to deal only with the relations of man to man. In reality, however, the question is what is his attitude to the world and all life that comes within his reach.
>
> (Albert Schweitzer, *Out of My Life and Thought*)

The central theme of this book is the practice of animal husbandry for the sustained good of all – humans, farm animals and the living environment. Previous chapters have examined the scientific principles that underpin the use of resources in animal farming systems, the health and welfare of the animals involved in these systems, and the quality and safety of food from animals. However science alone can do no more than explore what can be done and how. It contributes little or nothing to questions as to what should we do, what would we like to do and what can we afford. These are matters of philosophy, politics and economics (PPE). This chapter will explore the PPE of animal husbandry in the spirit of the above Schweitzer quotation; i.e. it will extend the conventional 'man-to man' nature of PPE to consider our attitudes and actions in regard to the whole living planet. It poses questions such as:

- What do we mean by good?
- Who and what are included within the definition of all?
- How effective and necessary is the law in delivery of the good?
- How effective is a moral belief in the good when set against the biological basis of individual human behaviour and the politics and economics of human societies?
- To what extent can human societies evolve towards the good with no help from moral philosophers?

This requires an examination of ethics as they determine good practice and the application of ethics through political and private actions within the context of the law, realism and sound economics.

Aims and opportunities for total agriculture were outlined in Chapter 1 (Table 1.2) as the production of food and other goods and services for human use, management of the land and investment in resources. Every opportunity for the individual was linked to a matching responsibility to society. This is the ethical core of the Social Contract, most beautifully expressed by Hobbes in his *Leviathan* (1651). He first described what he called a State of Nature (very close to the Darwinian concept of survival of the fittest) where:

> It is manifest that without a common power to keep them all in awe, men are in a condition of war of every man against every man. In such condition there is no place for industry, because the fruit thereof is uncertain: and consequently no culture of the earth; no navigation; no commodious building; no knowledge of the earth; no account of time; no arts; no letters; no society; and which is worst of all, continual fear, and danger of violent death; and the life of man, solitary, poor, nasty, brutish, and short.

Given that this was a bad thing, he argued that:

> In pure self-interest and for self-preservation men entered into a compact by which they agreed to surrender part of their natural freedom to an absolute ruler in order to preserve the rest. The State determines what is just and unjust, right and wrong; and the strong arm of the law provides the ultimate sanction for right conduct.
>
> The [common power] is called a COMMONWEALTH. This is the great LEVIATHAN, or that mortal god, to which we owe, under the immortal God, our peace and defence.

Modern convention would interpret Hobbes as saying that our quality of life as individuals depends upon our commitment to life in an ordered, fair and functional society, where each individual has a responsibility to contribute to the general welfare and accept some restriction on personal freedom. In most modern argument, the Leviathan or mortal God[1] does not require the presence of a just, but autocratic, philosopher/king but can be achieved through the just actions of those appointed, more or less democratically, to maintain justice within the commonwealth of nations. This principle is embodied with the Charter of the United Nations, which seeks (inter alia):

- to reaffirm faith in fundamental human rights, in the dignity and worth of the human person, in the equal rights of men and women and of nations large and small …
- to establish conditions under which justice and respect for the obligations arising from treaties and other sources of international law can be maintained …
- to promote social progress and better standards of life in larger freedom …

These are laudable aims, but it isn't easy, even when we interpret society only in terms of humans. My aim, of course, is to extend the concept of the commonwealth beyond the society of men (of all genders) to incorporate the welfare of all life on the land. You would be justified in dismissing this aim as a harmless but hopeless utopian dream were it not for the fact that ethical action is required now if we are to avoid a future that is ever nastier and more brutish. The definition of the Commonwealth has to extend beyond human society to include not only sentient farm animals but also the welfare of all life on the land (the biota). By defining the issues in terms of opportunities and commensurate responsibilities, I deliberately avoid use of the concept of 'rights'. Instead, I identify the importance of 'reward': reward for the producer who practises good husbandry; reward for the consumer who supports the practice of good husbandry; and reward for farm animals and the biota, who get a better deal.

Ethical principles

Ethics, synonymous with moral philosophy, is a structured approach to examining and understanding the moral life. Indeed, it can be called the science of morals. There are two classic approaches to addressing moral issues, conveniently abbreviated as 'top–down' and 'bottom–up' (Webster *et al.* 2010). The classical 'top–down' approach asks the question: 'Which general moral norms for the evaluation and guidance of conduct should we accept and why?' The principal aim of this approach is to justify moral norms. The drawback to this approach is that practical issues may be given little emphasis or ignored. The alternative 'bottom–up' approach is first to identify a specific practical issue, then construct an analysis of relevant moral issues by a process of induction. Beauchamp and Childress (1994) have developed a powerful and widely adopted 'bottom–up' approach to addressing problems in biomedical ethics. This approach has been developed in the context of agriculture and animal husbandry by Mepham (1996) to create an 'Ethical Matrix' built upon three aims of common morality: to promote 'wellbeing, autonomy and justice' as applied to all concerned parties, producers and consumers, farm animals and the biota. The first of these principles of common morality, particularly as it applies to the treatment of farm animals, is utilitarianism. This requires us to promote the well-being of the greatest number through the practice of beneficence ('do good') and non-maleficence ('do no harm'). The second principle of common morality, defined by Mepham as autonomy, is probably better defined as deontology, the science of duty, from the Greek *deon* (duty) and *logos* (science). Deontology, in this context, commands us to respect the 'autonomy' of each individual ('do as you would be done by'). This is consistent with the Buddhist principles of right thought and right action. The two principles define responsibilities but, standing alone, do not offer any reward, save that of knowing that we do the right thing (or 'thy will' for those with a God). The elements of responsibility to all concerned parties are what determine 'justice': i.e. equality and fairness whereby right actions should bring appropriate rewards, and wrong actions should carry appropriate penalties. Conventionally, justice has been defined as an ethical principle, i.e. an input to the ethical analysis. Having argued in Chapter 3 that the analysis of welfare is best achieved by assessment of inputs and outcomes, I suggest that justice is better considered as an outcome indicator of the extent to which practical application of the principles of well-being and autonomy (assessed from the input evidence) has succeeded in achieving justice to all concerned parties (assessed from the evidence of outcomes).

Application of the principles of utilitarianism and deontology to achieve the aims of well-being and autonomy within human society, considered in isolation, is relatively straightforward since we all have (or, in a just society, should have) opportunities and each of these opportunities carries a responsibility. When we introduce farm animals and the living environment into the ethical analysis, we complicate the problem since they cannot contribute to the debate. Humans are moral agents and carry moral responsibilities. The animals and the living environment are 'moral patients'. We have a responsibility to them but they have no responsibility to us. (In this sense, their position is similar to that of new-born babies.) The concept of justice with respect to the farm animals and the living environment demands that we, the moral agents, should seek a fair and humane compromise between what we take, in the form of food and other goods, and what we give, in terms of good husbandry; competent, humane and sustainable concern for all life that comes within our dominion.

Table 5.1 illustrates the Ethical Matrix in a form similar to that outlined by Mepham (1996) in which the columns are well-being (beneficence and non-malevolence), deontology and justice. Four rows define the concerned parties: producers and consumers (i.e. society at large), the moral agents; farm animals and the living environment, the moral patients. The phrases within each of the boxes are intended to do no more than provide brief introductions to the various arguments.

For the consumer (society at large) the utilitarian aim of beneficence implies the promotion of our general well-being through access to sufficient wholesome, safe, cheap food. The aim of autonomy implies that each of us should have the freedom (the right) to select the food we want, on the basis of personal taste, be that least cost or high welfare, even when it may not be consistent with our health in the long term. The duty of government and others in authority is to guide us towards educated and healthy choices, not to dictate what we eat. Most would also agree that the well-being of us, the people, is enhanced by contact with other living creatures and access to the countryside for physical and spiritual recreation. The well-being of producers and landowners depends on sufficient financial and other rewards to support a reasonable quality of life for themselves *and their animals*, and to enable them to take pride in their work. A clear illustration of this principle comes from a perennial grumble of the dairy farmer: 'I would love to improve the quality of my housing and the care I give to my cows but I don't have the money and I don't have the time.' The aim of autonomy implies that individual producers should be able to compete within a fair market, both at the national and international level.

Table 5.1 Principles of good husbandry: the ethical matrix

	Beneficence	*Autonomy*	*Justice*
Moral agents			
Human society at large	Wholesome, safe, cheap food Access to the countryside	Freedom of choice	Fair food pricing Legislation and incentives Production methods and land use
Producers and land owners	Financial reward Pride in work	Free competition	Fair trade Good husbandry
Moral patients			
Farm animals	Competent and humane husbandry	Environmental enrichment Individual freedom of choice	'A life worth living'
The living environment	Conservation Sustainability	Biodiversity 'Live and let live'	Respect for environment and stewards of the environment

The concepts of rights and responsibilities apply only to the moral agents. The right of society at large to demand food that is wholesome, safe and fairly priced carries responsibilities to all other concerned parties: the producers, the animals and the living environment. In practical terms, this means we must recognise (at least) the need for legislation to ensure acceptable standards of animal husbandry and environmental care. We also have a moral responsibility to encourage improvements to husbandry through incentives to improve upon the status quo. I shall develop this theme later under the banner of 'politics by other means'.

Justice for the producer balances the right to free trade and a decent income against the responsibility to practise good husbandry. According to the classic, Adam Smith free-market argument, farmers should have no special rights. They are just one group within the overall division of labour so should be served no better nor worse than any other group by the 'invisible hand of the market' (Smith 1776). I hesitate to criticise Adam Smith but I contend that farmers have a special case. We all carry a responsibility to care about sentient farm animals and the living environment. However, *caring about* and *caring for* are not the same thing. In developed, urban societies only a tiny minority are in direct contact with farm animals, or in a position to make an active contribution to the care of the environment. The landowners, farmers, herdsmen and shepherds are the *de facto* stewards: those with the direct responsibility of care. If it is our wish that they should do these things better then it is our responsibility to give them our support.

Delivery of justice to the moral patients requires us to reward the farm animals who provide us with food and clothing, work and financial security with the provision, at least, of 'a life worth living', to use the words of the UK Farm Animal Welfare Council. Justice for the living environment balances our right to enjoy it against our responsibility not to wreck it. The principle of beneficence for the farm animals requires us to promote the well-being of the flocks and herds through competent and humane husbandry. However, according to the principle of autonomy, this alone is not enough. It requires us to recognise the rights of individual animals to exercise some freedom of choice so that they can make a constructive contribution to the quality of their own existence. In practice, this can involve freedom of movement, environmental enrichment and some opportunities for diet selection. These issues were discussed in Chapter 3.

The utilitarian approach to the proper management of the living environment requires action to sustain the habitat and conserve the species contained within that habitat or living system. Wherever farm animals are part of the living system, the landowner dictates the nature and quality of the environment. An environment that may appear entirely satisfactory, indeed delightful, to the landowner and other humans who enjoy the land may be less than ideal in terms of the welfare of the wild fauna and flora.

The principle of deontology recognises our obligation to respect the needs of all members of the fauna and flora, even though they may make no contribution to our own well-being. We need to be realistic about this. It can do no harm to hug a tree but we shall not contribute to the overall quality of the environment, e.g. within an agro-forestry system, by seeking to preserve, for emotional reasons, an aged and rotten tree that is contributing nothing to carbon capture, shade for the cows or income for the overall enterprise. We cannot hope to do well by every individual plant and animal in the living environment, not least because, in most cases, we are unaware of their existence, inadequately aware of their needs, or because their needs conflict to a greater or lesser extent with ours. Nevertheless we should recognise that other creatures have needs, respect these needs and help the stewards of the living environment to attend to these needs.

The final, and perhaps the most important point to be made about the ethical matrix is that it is comprehensive. When we examine all the boxes, it forces us to consider our responsibilities to the people close to home and far away, the animals both domestic and wild, the life of the land and the stewards of the land. It addresses one of the central aims of this book, in that it renders single-issue arguments impossible. Those who call for the imposition of gold standards for animal welfare and can pay for the produce have a responsibility to consider those who cannot. Those who want to slaughter badgers to protect the dairy industry and those who want to preserve all badgers whatever the cost to the industry must not shut their ears to the arguments of their opponents. A just outcome requires compromise between all stakeholders: moral agents and moral patients. We, the moral agents, have the responsibility to acknowledge that it is not fair to expect all we would like. We may derive satisfaction from observing that our moral patients are faring well but we cannot expect any gratitude.

Animal husbandry and the law

The primary function of the law within society (national or international) is to define what actions are and are not acceptable in order to 'establish conditions under which justice and respect for obligations can be maintained'. This makes it the core, although not the entire substance, of Hobbes's Leviathan. It is helpful to make a distinction between proscriptive laws, regulations, codes of practice and legislation to create incentives, e.g. through the use of subsidies. Proscriptive law seeks to define that which must be done, must not be done or not left undone. Regulations seek to explain the details of the law, the reasons for the law, give directions on how to meet the formal requirements of law, and the principles of justice that underpin the law. Codes of practice, such as the DEFRA codes of practice for the welfare of farm animals (or the UK Highway Code), do not carry the force of law. However failure to abide by codes of practice can be used as evidence of lawbreaking, e.g. failure of an owner to maintain a duty of care to his animals.

Proscriptive law exists to meet the needs of society to give protection (to humans and animals) from lives that are 'solitary, poor, nasty, brutish, and short'. The implementation and interpretation of the law are matters for the judiciary, but the laws are written by politicians and, in a democracy, politics is expected to adapt and evolve to reflect the changing perceptions of just people for a just society. Think slavery, votes for women, battery hens. However, while proscriptive law and regulations do evolve in accord with changing standards of acceptability, they can do no more than enforce standards of acceptability operating at the time. Legislation by incentive (e.g. subsidies) is designed to encourage people to do better: to set standards that are higher than merely acceptable.

A short history of the law in regard to the protection and welfare of animals in the UK provides a pioneering and useful example of the evolutionary process (Radford 2001). Prior to 1822, animals were defined as property and had no legal protection in their own right. In 1822 the UK Parliament passed 'Martin's Act' to prevent the cruel and improper treatment of cattle, making this the first parliamentary legislation for animal welfare in the world. The 1835 Protection of Animals Act made illegal certain deliberate acts of cruelty to captive animals: e.g. bull, bear and badger baiting, cock and dog fighting. This legislation did not extend to wild animals.

The most substantial piece of UK legislation for the welfare of domestic animals has been the Protection of Animals Act 1911. This established the concept of 'causing unnecessary suffering by doing, or omitting to do, any act'. This law has stood the test of time because it is both comprehensive and robust, although it may be argued that in practice it has been applied almost entirely to sins of commission. The recent UK Animal Welfare Act (2006) has imposed a legal duty of care on responsible persons to provide for the basic needs of their animals (both farm animals and pets). Under this Act it is not necessary to prove that suffering has occurred; it should only be necessary to establish that animals are being kept, or bred, in a way that is liable to cause suffering. This new Act has yet to be fully road-tested by case law. It should, for example, protect broiler chickens from the probability of suffering foot lesions through inadequate provision of bedding. As it reads, it should also protect broiler chickens from a high prevalence of painful leg disorders as a direct consequence of breeding policy. To date no one has put this interpretation of the 2006 Act to the test. It will require a lot of courage – and some good lawyers.

The World Association for Animal Health (OIE, Office International Epizootique) has set down international laws, regulations and codes of practice in relation to the husbandry, health and welfare of farm animals. These include the *Terrestrial Animal Health Code* (modified 2011), which deals mainly with 'veterinary' hazards and controls in relation to (e.g.) infectious diseases, semen collection and certification for international transport. It does consider animal welfare in relation to specific issues such as transport and slaughter but, at present, does not address the quality of animal husbandry on farms.

Proscriptive laws tend to be worded in very broad terms, and this is wise. It is a sound and lasting principle that one should not cause 'unnecessary suffering' to a sentient animal by doing, or omitting to do any act. The main reason why the concept of unnecessary suffering is robust is that it is flexible. It can accommodate developing concepts within and between societies in the meaning of the words suffering, unnecessary and sentience. This is most visible in the context of the law dealing with the treatment of animals used in science. In the Cruelty to Animals Act 1896, which sought to regulate experiments with laboratory animals, suffering was equated with pain and the necessity of the procedure was decided simply on a yes or no basis. The Animals (Scientific Procedures) Act 1986 broadened the categorisation of cost to the animals to include pain, suffering, (emotional) distress and lasting harm, ranked cost severity as mild, moderate or substantial and required all scientific procedures carried out under the Act to justify the likely costs (harms) to the animals in terms of likely benefits to society (humans and other animals). Within the EU definition of animals included within the Act, by virtue of assumed sentience, animal, which formerly meant vertebrates, has been extended to include cephalopods (e.g. *Octopus vulgaris*). There have been similar developments in the legal interpretation of the concept of 'unnecessary suffering' in the context of farm animals. Examples include new regulations for the keeping of breeding sows and laying hens. Barren cages for laying hens, and individual confinement stalls and tethers for pregnant sows are now banned (or about to be banned) throughout the European Community. As yet there is no general ban within the USA. However a number of individual states (six at the time of writing) have imposed local bans and the number is rising. In Australia the pig industry, supported by some tame vets, have been vociferously opposed to a ban on sow stalls. This resistance has now been broken by the power of the people. Individual supermarkets, acting in response to public opinion, refused to buy pork from units employing sow stalls and the industry capitulated in short order.

It is undeniable that developments in national and international laws and regulations have controlled the worst forms of exploitation of land and labour (people and animals) and contributed significantly to improvements in food safety, animal health and husbandry. However, the explosion of demand for free-range eggs within the UK and the impact of the supermarkets on the Australian pig industry illustrate the fact that progress towards standards that are higher than the merely acceptable is likely to be faster when driven by responses to consumer pressure in a highly competitive free market than by the leisurely implementation of incentives by government establishments strapped for cash. Nevertheless it is worth quoting Hart (1961):

> It cannot seriously be disputed that the development of the law . . . has been profoundly influenced both by conventional morality and ideals of particular social groups, and also by forms of enlightened moral criticism urged by individuals whose moral horizon has transcended the morality currently accepted.

The law may never be enough, but it is always essential. The needs for the law in relation to our treatment of farm animals have been succinctly expressed by Radford (2001):

1. The law sets absolute standards for animal welfare not amenable to whims of producers or retailers.
2. The law is all embracing: it does not permit individuals or groups to opt in or opt out.
3. The law makes those who use animals accountable for their care.
4. The law ensures that society as a whole has responsibility to promote animal welfare.
5. The law reflects and adapts to the collective concerns of society.
6. The law takes account of non-commercial concerns that are outside the market: e.g. wild animals and the living environment.
7. Belief in the principles of the law conveys a sense of trust without the necessity for everyone to examine all regulations line-by-line.

Regulations

For the law, expressed in broad terms, and based on basic principles of justice, to operate in practice, it needs to be backed up by regulations which address in detail what should be done, to whom it should be done and why. The European Commission has prepared a series of regulations in relation to the protection of animals. These comprehensive (and lengthy) regulations are made up of the following elements:

- terms of reference, background and reasons for the regulations;
- scope of the regulations and necessary definitions;

- conditions that must be fulfilled to meet the regulations;
- operation of the regulations.

An annex then sets out the technical rules for operation within the regulations.

A useful and important illustration of these regulations can be obtained from examination of the European Council Regulations on the Protection of Animals in Transport (EC1/2005). Box 5.1 summarises the major conditions of this regulation. The general principle of the law requires that no person shall transport animals in a way likely to cause injury or undue suffering. The conditions necessary for this include (inter alia) that the animals are fit to travel, loading facilities and the environment within the vehicle are satisfactory, the handlers are competent and humane, transport is carried out without delay, animals are inspected and given feed and water as necessary. Annex 1 to EC1/2005: Technical Rules is, in essence, an instruction manual giving precise directions for space allowances, journey times, rest intervals, feeding and watering, etc.

The aim of regulations is to give precise, helpful and essential directions to enable people to operate within the law. These regulations are regularly reviewed and amended as necessary in response to advice, mostly from scientists in the light of new scientific evidence. This is healthy, although in my experience these reviews would benefit from better communication between academics and those involved in the business (e.g. animal transport) on a day-to-day basis. There is, moreover, always a danger of overregulation. Nevertheless, regulations are essential particularly when they relate to issues as important as animal suffering. When drafting regulations, the aim should be to strike a balance between carrot and stick, while avoiding pettifogging intrusions on personal liberties and lengthy expositions of the blindingly obvious.

Regulation by incentives: subsidies and regulated markets

The first function of laws and regulations respectively is to control behaviour by prescribing in general and in detail what individuals in a society should and should not do. A second function is to create incentives (usually financial incentives) to encourage and reward individual behaviour that is consistent with the aims of society at large. The Common Agricultural Policy (CAP) within the European Union was established as a major instrument for regulating land use and the market for agricultural goods within the European Community. In 2010 expenditure on the CAP was €372 billion. While the CAP has evolved over time it was always intended to contain elements of both carrot and stick. Initially it was designed to regulate supply and demand for goods through subsidies linked to production, quotas and set-aside. In 2003, at a time of plenty, EU farm ministers adopted a fundamental reform of the CAP, based on 'decoupling' subsidies from specific products. Payment for production was replaced by a new 'single farm payment', subject to 'cross-compliance' conditions relating to environmental, food safety and animal welfare standards (EC 2012). This acknowledged the need to reward farmers not only for their production of commodities but for

Box 5.1 Extract from regulations on the protection of animals in transport (EC 1/20005)

No person shall transport animals or cause animals to be transported in a way likely to cause injury or undue suffering to them.

The following conditions shall be complied with:

(a) all necessary arrangements have been made in advance to minimise the length of the journey and meet animals' needs during the journey;
(b) the animals are fit for the journey;
(c) the means of transport are designed, constructed, maintained and operated so as to avoid injury and suffering and ensure the safety of the animals;
(d) the loading and unloading facilities are adequately designed, constructed, maintained and operated so as to avoid injury and suffering and ensure the safety of the animals;
(e) the personnel handling animals are trained or competent as appropriate for this purpose and carry out their tasks without using violence or any method likely to cause unnecessary fear, injury or suffering;
(f) the transport is carried out without delay to the place of destination and the welfare conditions of the animals are regularly checked and appropriately maintained;
(g) sufficient floor area and height is provided for the animals, appropriate to their size and the intended journey;
(h) water, feed and rest are offered to the animals at suitable intervals and are appropriate in quality and quantity to their species and size.

their commitment to total agriculture; the sustained husbandry of the land and the life of the land.

Calculation of the amount of the single farm payment has been based on two pillars. Pillar 1 involves direct payments to farmers. Pillar 2 is more broadly available for projects linked to rural development and includes issues such as support for less favoured areas, agro-forestry, agro-environment programmes (including water management) and incentives to tourism and local craft industries. At the start of this policy in 2003 the proportion of subsidy paid directly to farmers under Pillar 1 was 89 per cent. Direct payments under Pillar 1 have been closely linked to production and farm size, within quotas as appropriate. Receipt of this subsidy is conditional on cross-compliance with defined standards that include environmental protection and animal welfare. There are penalties for non-compliance, starting at 3 per cent in the first instance but in theory rising to 100 per cent in cases of 'wilful and persistent non-compliance'. Farms are inspected annually for compliance but the evidence suggests that nearly all farms pass muster, which implies that so far the policy is doing little to discriminate between farms on the basis of non-production values. Thus by far the greatest proportion of subsidy through the CAP has been going to farmers simply because they are there. The CAP single payment policy that doles out money to farmers simply in proportion to the size of their farms, or even, in less favoured areas (e.g. the Scottish Highlands) on the basis of 'headage' (the numbers of animals on the farm at the start of the scheme) has been widely criticised on the basis that it creates no incentives to do the right thing, does not reduce the cost of food to the people and simply gets swallowed up in increased land prices.

In 2010 the European Commissioner for the Environment Janez Potočnik called for a *Common Agricultural and Environmental Policy*, saying that 'the CAP should be greened; that it should improve sustainability, soil quality, water quality and efficiency. The policy should contribute to *global* food security and provide green products.' A revision of the CAP in 2011 proposes that, in future, 30 per cent of the direct payments 'will be made contingent on a range of environmentally-sound practices, going beyond cross-compliance'. There is as yet no proposal to reward standards of animal welfare that 'go beyond cross-compliance'. Some of the environmental incentives within the new Pillar 1 will be offset by a serious reduction in the already small amount of money available from Pillar 2. In 2013 this will be only €15 million, or 4 per cent of total expenditure. The proposal to green the CAP in order to reward environmental stewardship properly acknowledges the responsibility of society as a whole to support and direct the stewards (the farmers) to sustain the land in a way that will continue to meet the needs of succeeding generations of society (the moral agents) and the biota (the moral patients). The question is: 'Has it gone far enough?'

In regard to farm animal welfare, the CAP at present demands no more than that farmers should meet nationally acceptable standards. It does not directly reward 'high welfare' systems. This has been criticised by some welfare charities. However the ethical matrix would suggest that consumers who demand higher than acceptable standards of animal welfare when they buy food should be prepared to pay for it themselves.

My personal conclusion is that the new green CAP can be a powerful aid to the practice of 'planet husbandry', not least because it represents a lot of money. In time one would wish to see it evolve into a system whereby subsidy is more precisely linked to the specific actions for the good of the environment (e.g. C sequestration, extension of wildlife habitat including the provision of corridors between reserves). I shall return to this issue in the final chapter.

The economics of animal husbandry

Economics is best considered as being amoral. It is an examination of the real world that stands alongside but entirely separate from politics and philosophy. Its scientific credentials are based on the fact that it seeks to avoid value judgements in the observation, testing and interpretation of human behaviour.

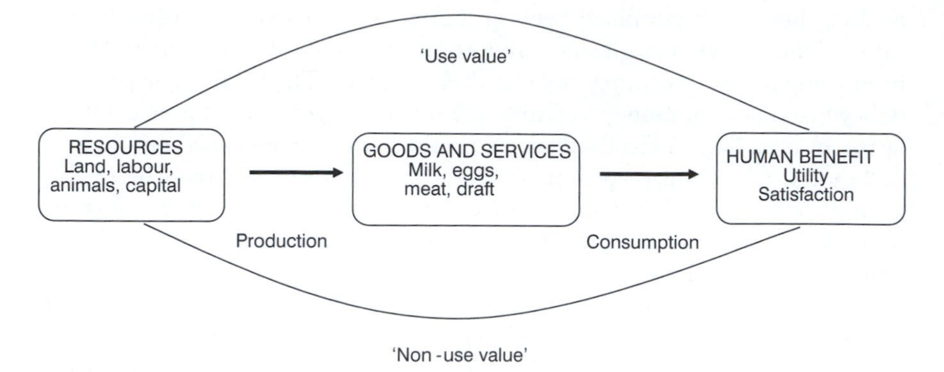

Figure 5.1 The economics of animal husbandry (from McInerney 2012)

The economics of animal husbandry and welfare has been excellently reviewed by John McInerney (2012), and what follows is largely based on his analysis (see Figure 5.1).

Economics involves an analysis of the processes by which finite means are used to meet competing ends. The means are defined as *resources*. In the context of animal farming, these are land, labour, animals and capital. The ends that drive the use of these resources are *human benefits*. Resources yield human benefit by being converted into *goods and services* that, in a free market, compete with one another to meet human desires for benefits, which are defined in Figure 5.1 as utility and satisfaction. Provision of sufficient food to support life is a utility, an absolute demand. However, wherever people have access to a choice of foods and the means to express their choice, food is both a utility (a source of essential nutrients) and a major source of satisfaction. The more affluent the society, the greater the ratio of satisfaction to utility and the greater the probability that food consumption and food selection will exceed our biological needs. The value of goods and services measured in terms of utility and satisfaction are defined, respectively, as *use value* and *non-use value.*

Food items

The utility value of a food product on sale in competition with other products is defined by its nutritive value, safety, availability/scarcity and convenience. Everything else that contributes to its value (measured by the price we choose to pay) is a non-use value measured in terms of the satisfaction it gives us. The nutritive value of caviar and fish paste is much the same. As I write, fish paste is retailing at about 45p/100g, caviar at £25–£80/100g. To an economist, the criteria that individual consumers recruit to determine the non-use value they give to the benefit they derive from a food product can all be measured on the same scale. These may include appearance, anticipated taste, provenance (nation, region or farm of origin), application of fair trade principles, production method (e.g.

organic, Freedom Food, wildlife sensitive) and fitness to our own perceptions of style and status (the phrase 'reassuringly expensive' has proved to be an effective marketing slogan for lager). Retailers who operate in the free market and profit from providing food to suit all tastes are content to accommodate any or all of these perceptions of benefit and recognise no economic distinction between adding value through the provision of Beluga caviar, Aberdeen Angus beef, organic pork or high-welfare free-range eggs.

The extent to which these non-use measures of value increase the price of the product determines the reward that producers can obtain in respect to the goods they produce and the means of production. The latter include regard for fair trade, animal welfare, organic principles, sustainability, wildlife conservation (etc.). Thus while the unregulated free market operating entirely according to economic rules may be amoral, most of the drivers of the market, especially for the non-use measure of the value of food products, are driven by individual value judgements and many of these may properly be called moral judgements. These are measures of the extent to which consumer beliefs drive production methods and the consequences thereof for all concerned parties, including the moral patients, farm animals and the living environment.

Since all these non-use elements of value can be classed together in economic terms, they can be considered as additive. Thus a particularly discerning consumer might be prepared to pay a very high price for a steak from a pure-bred Aberdeen Angus steer, reared at pasture on an organic farm to an age in excess of 30 months to Freedom Food high-welfare standards, slaughtered on farm and the carcass matured for four weeks at 4°C. The relative magnitude of each of these components of added value will, of course, vary with the consumer. The global economic downturn in the period beginning 2008 has, in the UK, led to a significant reduction in demand for organic foods, whereas sales of allegedly high-welfare free-range eggs has remained stable. It would be tempting to conclude from this that public demand for high-welfare food is less fickle than the demand for organic. It is however necessary to point out that organically produced eggs currently cost about 60 per cent more than eggs from caged birds, whereas (when I last looked) the price difference for Freedom Food, non-organic free-range eggs vs. cage eggs was about 10 per cent. One cannot, at this stage, conclude whether or not there is a real difference in the flexibility of economic demand for high welfare vs. organic food.

One fact of economics and marketing is that the returns from marketing product A on the basis that it is even better than another good product B will always be much less than the returns from marketing product A on the valid or invalid basis that product B is actually bad. The conspicuous increase in the sale of free-range eggs relative to other lines of advertised high-welfare foods is a conspicuous example of this rule. The greatest influence on public opinion has undoubtedly been the image of the battered hen in the battery cage. It becomes less easy to advocate the case for higher welfare standards for the dairy cow when the image, heavily promoted by the industry, is one of happy cows in green fields, even in countries (e.g. USA) where this is the exception to the rule.

The amoral economic world does not require cows to be in green fields, it only requires us to believe that cows are in green fields.

It would be unfair, wrong and probably illegal to claim added value for 'higher welfare' food items on the basis that conventional production methods, acceptable according to law, regulations and codes of practice, are bad. It is fair to argue that, if things can be done better, there should be higher rewards, but progress down this honourable route is likely to be slow.

There is however one element of the amoral free market that works in favour of farmers and their animals. Demonstrable concern for animal welfare is recognised by all supermarkets as a key element of their corporate image, and promoted as a powerful incentive for you to enter their shop because, in this regard, this makes them more loveable. Once in the supermarket, of course, the shopper is likely to spend a small proportion of their weekly budget on high-value, high-welfare goods, then fill up their trolleys with the same beer, beans and crisps that they could buy anywhere.

Non-food items

A central principle of this book is that farm animals cannot be considered simply as resources for the production of food as a commodity. Whether viewed on a global scale or at farm level, a comprehensive audit of the value (use and non-use) of the elements of total agriculture should include the following:

- Utility value
  - for individuals: food, draft, clothing, fertiliser, capital;
  - for society: sustainability of rural communities;
  - for the living environment: management of C and N cycles.
- Satisfaction value
  - for individuals: animal welfare (human perception);
  - for society: habitats for wildlife, amenity, beauty;
  - for farm animals: animal welfare (animal perception).

The utility values of farm animals are self-evident: food, draft, clothing, fertiliser, capital. The relative economic importance of these different measures of value differs between societies and changes with time. The value of a yak to a Nepalese family and the value of a Holstein cow to a large commercial dairy producer in the USA are measured in very different ways. The value of wool in the Middle Ages brought great wealth to the permanent pastures of Great Britain. Now, in the same environment, it does not repay the cost of shearing. The utility value for society of mixed farming systems is measured in terms of their contribution to the quality and sustainability of rural communities. Politically (e.g. within the CAP) this is referred to as 'sustained rural development'. The word development is, of course, political newspeak for 'reduced rate of decline'. The third utility value is measured in terms of the physical impact of the agricultural system on the quality and sustainability of the physical and biological environment. This can, of course, be positive or negative, which means in economic terms that it should be addressed with both carrot and stick. It is right that the polluter should pay. It is equally right that good husbandry, whether measured in terms of energy conservation, water management or carbon sequestration, should be rewarded.

The satisfaction (non-use) value of farm animals and their products to individuals can be measured in terms of the value we give to provenance, production methods and animal welfare. It is an inescapable fact of life that individuals will differ in the added value they give to these things, whatever may be the minimal standards set down by legislation. Satisfaction for society at large comes from such things as the value we give to the quality of the living environment, *as perceived by us*, using measures such as wildlife numbers and diversity, amenity for recreations ranging from hiking to hunting, and a general respect and craving for beauty. For the animals themselves the satisfaction value of any husbandry system has to be measured in terms of their own welfare. Have their lives been good, worth living or not worth living? In any analysis of planet husbandry, political, philosophical, economical, or all three, all these elements of agriculture need to be taken into account.

Rewards

Any system for the production of goods, services and amenities within a society that has the freedom of choice will only succeed if the rewards are commensurate with the costs. In a largely free-market society these rewards are mostly immediate or short-term and apply only to the human population of consumers and producers. Individual consumers select products on a cost–benefit basis. The simplest measure of this is price per unit product. In the case of food from animals, consumer measures of value are complex and include appearance, taste, food safety, provenance and animal welfare (inter alia). Some consumers will derive reward from buying eggs 10 per cent cheaper from Shop A than Shop B. Others will derive reward from buying eggs that are certified in the shop as free-range or purchased direct from the farmer. Consumer perception of benefit is part of the overall analysis of reward. However consumers vary greatly in the way they value reward, i.e. there is much more to consumer reward than value for money.

The most obvious measure of reward to the producer is profit margin, measured in terms of income and expenditure. This can be measured in two ways. *Gross profit margin* sets income from sales (e.g. meat, milk, eggs) against *variable costs* for feed, bedding, vet bills, etc. *Net profit margin* includes fixed costs (e.g. labour, interest on loans, amortisation of buildings and equipment). The latter is ultimately what determines success or failure. However, in the family farm, whether in the developed or less developed world, the value of labour is not measured simply in cash terms. I illustrate this with a touching example. Some years ago I was at a national (UK) dairy event giving advice on the husbandry of artificially reared dairy calves. On sale at the event were automated nipple feeders that dispensed measured quantities of milk replacer to calves, as an alternative to bucket feeding. In strictly cash terms I estimated that it would need to feed about 30 calves at any one time to justify the cost. One farmer, on a small unit with seldom more than six calves on milk at any one time, chose to buy one as a silver wedding anniversary present for his wife to relieve her from the daily chore of bucket feeding calves that had occupied her for the last 25 years.

The income of many farmers, assessed per hour worked, can be pitifully small relative to the take-home pay of most skilled workers in industry (and this applies world-wide). This begs the question why so many farmers are prepared to work such long hours for so little financial reward. One of the sadder reasons is that many are tied to their land and their animals and they cannot escape. However, it is also undeniably true that many farmers have a robust sense of duty to their herds and the land for which they are responsible and which they would wish to pass to successive generations. This can certainly be called husbandry. It is not too much to call it love. It is obviously easier to love farm work when it is carried out on a relatively small scale that permits regular direct contact with the animals and with the land than as an operative in an industrial unit for caged laying hens with ten buildings each containing 30,000 birds. The Bristol welfare and behaviour group carried out an evaluation of the welfare of hens on free-range units operating under the aegis of Freedom Foods. In general the welfare of these hens was good (Whay *et al.* 2007). One of the most exciting things about this study was the farmers I met who had switched from rearing hens in cages to hens on free range and had recovered their enthusiasm for farming. Their income was neither better nor worse than before but their rewards were greater.

The principle of justice requires that rewards should also accrue to the moral patients: farm animals and the living environment (Table 5.1). Since they cannot express the extent to which they feel they have been rewarded, still less conclude whether the rewards are fair, we have the responsibility to pass justice on their behalf. The practical application of this justice is discussed in later chapters. It should include regular monitoring and surveillance of animal welfare and environmental quality within audit schemes that reward the farmers in proportion to the quality of their stewardship.

'Contract for life'

At the beginning of this chapter I introduced the concept of the 'Social Contract' defined by Hobbes as one where 'In pure self-interest and for self-preservation men entered into a compact by which they agreed to surrender part of their natural freedom . . . to preserve the rest'. This implies, of course, a series of more or less conscious decisions whereby individual humans within an ordered society concede that some sacrifice of personal freedom is necessary to ensure a quiet life. As I have argued above, it is not possible to extend the social contract, as defined by Hobbes, to animal husbandry, since we cannot infer that farm animals, still less the wild animals and non-sentient fauna and flora, have decided that it is in their interests to go along with farming practices as they stand. Nevertheless it is valid to argue that, in the early stages of domestication, certain species like dogs, horses, cattle, pigs and hens recognised advantages of interacting with human communities, including the opportunity to scavenge for scraps of feed and increased security from traditional predators. Humans have been wise enough to disguise the threat we pose to domestic animals until the very end.

While it would be wrong to suggest that there can ever be a real contract for life between humans and the rest of the living environment, we humans cannot neglect our responsibilities in these matters. As I wrote at the very beginning, these are not just moral issues, they will determine the future quality of all our lives. Currently there are massive and immoral variations in quality of life, bankers in the City of London vs. garment workers in Pakistan, hens on free range vs. hens in battery cages. There is no moral reason why those who can, should not enjoy a good life so long as it is not unduly parasitic on the rest of human society and unduly profligate in matters of environmental resources. There is an unavoidable moral case that says that other individuals and other societies should be given opportunities to catch up. This can only come about when resources of food, energy, people and other animals are subject to better husbandry and treated with more respect.

General conclusions to Part I: engaging with the problems

This brings to an end Part I. The aim has been to consider animal husbandry not in isolation but as an essential component of total, sustainable agriculture, considered in the broadest possible terms as 'planet husbandry'. The method has been to identify and incorporate as many components as possible within the analysis so as to avoid the dangers of arguing from incomplete premises. No attempt has been made to present a comprehensive review of all issues within this enormous subject. Instead, major issues like competition between food for humans and feed for animals, animal welfare and production diseases, methane emissions and carbon sequestration have been selected and illustrated by examples to demonstrate fundamental principles that should be applied to the investigation and analysis. The title of Part I, 'Engaging with the problems', is intended to convey the message that my intentions are entirely constructive. My remit is not to carry out an exposé of deficiencies in current animal husbandry systems but an identification, analysis, treatment and resolution of existing problems. In this section, which deals with analysis, I have tried to avoid personal bias in my examination of the evidence. I accept an occasional criticism and concede that my style is, at times, too personal, but that's the way I write. However I believe that at no time in Part I have I allowed my personal likes and dislikes to intrude upon the analysis. When I write that a system such as individual stalls for pregnant sows has been banned in Europe and many states of the USA, that is not my opinion, it is a matter of record. The heading to Part II, 'Embarking on solutions', is also intended to convey a constructive message. Approaches to the resolution of current and future problems will again be based wherever possible on impartial evaluation of the existing evidence. However, since I shall be dealing with the future, I have no option but to resort, on occasion, to surmise and speculation.

Part I has dealt with a wide range of issues and illustrated them with a large number of examples, too many to be condensed into a short summary. Nevertheless I would like to highlight four big themes within my overall argument and ask you to carry them forward into your reading of Part II.

Nearly all arguments are made from incomplete premises

This applies obviously to most 'scientific' audits of agricultural systems; e.g. use of plant crops for animal feeds without distinction between the competitive and complementary, criticism of greenhouse gas emissions from ruminants without considering C sequestration within pastoral systems. It also applies to 'ethical' critiques of animal husbandry systems based exclusively on assessment of animal welfare. This can be even worse when the assessment of animal welfare is itself selective, e.g. biased in favour of production measures (Freedoms 1–3), or behavioural measures (Freedoms 4–5), to suit the prejudices of the advocate.

The value of farm animals cannot be measured simply in terms of food production

In traditional agricultural systems the value of animals is measured in many ways: food and clothing, draft power, fuel and fertiliser, capital. In the most intensive systems for pigs, poultry and dairy cattle, the sale of meat, eggs and milk constitutes by far the largest source of income. However, the sentient farm animals committed to the production of these goods cannot (in European law) be treated simply as commodities. It follows that measures of the value of farm animals and goods from farm animals should incorporate measures of the quality of the production system (e.g. resources, hygiene and management) and the quality of life for the sentient animals within the system.

In many extensive and pastoral systems the income from the sale of food alone is seldom sufficient to provide adequately for the welfare of the farmers or their animals. This is a serious problem both in the developing world, where incomes are generally low, and in the developed world, where land prices are generally high. While many livestock systems, especially intensive systems, can be a major source of environmental pollution, extensive, pastoral and agri-forestry systems can play a critical role in the sustainable management, cycling and sequestration of material resources (e.g. water, nitrogen, greenhouse gases). If the polluter should pay, the cleaner should be rewarded.

The law is essential but the law is not enough

Proscriptive laws relating to the husbandry and welfare of farm animals are essential to define and enforce minimum standards of acceptability – which change with time. Progress above and beyond minimum standards of acceptability in matters of husbandry and welfare can, in time, be incorporated into new or revised legislation. However, in practice, the free market, responding to moral criticism and societal demand, can do it quicker.

Right actions need rewards

Significant action by the moral agents, farmers and consumers, for the moral patients, farm animals and the living environment, will only come about if it brings significant reward. This does not necessarily equate with increased profit. Sufficient profit accompanied by increased job satisfaction can be sufficient reward but no farmer can reasonably be expected to lose money on the deal. This requires that all significant contributions to the husbandry of the animals and the land should be included in assessing the rewards due to the stewards. The logic of this argument says that intensive (factory) farms whose sole output is the production of food as a commodity cannot expect any source of income other than that from food production (i.e. no subsidies). Moreover they should pay for any significant contribution to pollution.

Farms that can be shown to produce added-value food on the basis of (e.g.) animal welfare standards that are higher than the legally acceptable minimum, have the right to sell it at a higher price. However it can be argued that this higher price should be paid by the individual consumers who recognise these measures of added value, rather than imposed (e.g. through subsidies) on society as a whole.

Expenditure on improving the stewardship of the living environment is action for the lasting benefit of all life forms. Some of the costs can be recouped from individuals in recognition of increased access to country amenities. However it is not reasonable to expect most of the costs of conserving the living environment to be borne by a few concerned individuals, although the role of environmental and wildlife charities is significant and impressive. This is a responsibility for society at large. In practice, this equates to public policy for the allocation of tax revenues, e.g. in the form of 'green' subsidies. Practical ways and means for addressing all these issues will be discussed in the concluding chapter to Part II.

Note

1 The phrase 'under the immortal God' is unnecessary to the argument.

Part II

Embarking on solutions

6 Better, kinder food

> Thou mayest give them meat in due season.
>
> (Psalm 104)

The subtitle of this book is 'The Place of Farm Animals in Sustainable Agriculture', where agriculture is considered in its broadest sense to embrace not only food production but all the benefits that can be linked to the practice of animal husbandry. The story so far: Part I, 'Engaging with the problems', has been an evidence-based examination and analysis of current practice in animal husbandry. Part II, 'Embarking on solutions', moves to prediction and advocacy; how we might and should do things better in future. While this will inevitably involve some speculation, I shall try to ensure that the arguments do not stray too far from the evidence and what may rationally be deduced from the evidence. Inevitably that means starting from where we are now, however unsatisfactory this may appear to some.

The main aim of animal husbandry will always be the production of food for human consumption. The main aim of good husbandry within the context of total agriculture must therefore be the production of good, kind food within a sustainable and ecologically sound programme of sympathetic land management that may, to a greater or lesser extent, embrace the generation of energy from biomass, biofuels or wind farms, the use of pasture and forestry as carbon sinks, and a profitable role for animal husbandry in the management of amenity, wildlife and conservation areas.

The relative contribution of food production and other forms of land use to the overall enterprise will depend on the nature and value of the land and other natural resources, and the nature and value to consumers and to landowners of the different benefits to be derived from working the land. For the purposes of discussion we may identify four categories of husbandry system:

- Intensive livestock production systems involving considerable capital investment for housing and machinery and directed almost exclusively to the production of food as a commodity.

- Pastoral systems where livestock are the most ecologically appropriate source of food production but which offer scope for alternative rewards for good husbandry (e.g. C sequestration, wildlife conservation, etc.).
- Traditional, 'village', mixed farming systems in underdeveloped regions.
- 'Added-value' systems (e.g. organic, high welfare farming) in the more affluent countryside.

Within each of these systems, I shall, in the succeeding chapters, examine pathways to progress through:

- implementation of best practice in the light of existing knowledge;
- developments in science and technology;
- laws, regulations and government based incentives;
- people power and the free market.

Measures of quality

Since the first aim of good husbandry should be the production of good, kind food, the aim for the future must be 'better, kinder food'. The words good and kind are broad in scope: good for the consumer, good for the farmer, kind for the animals and kind for the living environment. A more comprehensive list of the criteria by which food from animals may be defined as good and kind is presented (in no particular order) in Box 6.1. The criteria necessary for evaluation extend beyond measures of the value of food to us as consumers to embrace the broad definitions of total sustainable agriculture, 'planet husbandry'.

All food for sale, whether in the African village market or the European supermarket, should be *wholesome*. Wholesomeness is a measure of quality relating mainly to food safety and defined by such things as composition, standards of hygiene in preparation and at the point of display, freedom from potential pathogens, food spoilage bacteria and other harmful substances such as toxic minerals, pesticide and antibiotic residues. These things are governed by food standards and food hygiene regulations that set out thresholds of acceptability for concentrations of potentially harmful substances.

Food should be *affordable*. For affluent, already well-fed consumers, affordability is measured against both use and non-use estimates of personal benefit. When shoppers deliberate whether to buy a joint of beef or a broiler chicken, cage, free-range or organic eggs, they are performing a quite complex exercise in mental arithmetic; weighing their personal estimate of value, measured in terms of perceived taste, health, provenance and production standards, against the price. For poor, underfed people in the third world the cost–benefit equation is much simpler and dominated by nutritive value: 'Can I afford to buy milk and eggs for my growing child?' There is a third group, increasingly common, especially in the developed world, namely the poor but fat. The biological and social origins of obesity are complex and outwith the scope of this book. However we should acknowledge that many fattening foods

Box 6.1 Criteria necessary for the evaluation of food from animals

Part 1

Wholesome

Quality as defined by composition, hygiene, safety and freedom from potential pathogens, toxic agents and other harmful substances.

Affordable

Cheap enough to be eaten in sufficient quantities to meet specific nutrient requirements (e.g. amino acids from high-quality protein).

Attractive

Visually attractive, tasty and appealing to social standards of dining as entertainment and social intercourse.

Part 2

Healthy

Consistent with nutritional needs when eaten as part of a mixed diet in circumstances ranging from poverty to affluence.
Low-risk in relation to food and health (e.g. calories, saturated fats).

Part 3

Humane

Produced within systems where the resources and management are compliant with, or improve upon, national and international standards of animal health and welfare.

Efficient

Produced within systems that are efficient in terms of inputs (competitive and complementary feeds, fossil fuels, land use and capital investment).

Economic

Produced at a price that is fair to all stakeholders in the food chain: producers, retailers, consumers. This price will vary according to the value that individual consumers give to 'non-use' elements of benefit.

Ecologically sound

Produced within systems that are free from pollution (e.g. N and P), minimise *net* greenhouse gas production, preserve soil and water courses, conserve and create habitats for wildlife.

are also cheap foods and it is very difficult to fill your basket with 'healthy-living' products when you are operating on a very small budget.

Food should be *healthy*. Here the association between food and health is considered in a nutritional context. The risk from pathogens and other harmful substances (i.e. food safety) comes within the definition of wholesomeness. This subject was considered in more detail in Chapter 4. The key issue to be repeated here is that, while it is proper to describe an overall diet (measured in terms of quality and quantity) as healthy or unhealthy, it is seldom sensible to apply the terms healthy and unhealthy to individual foods. A light salad lunch may be a healthy choice for a fat banker but far from ideal for a starving African mother unable to make enough milk for her baby. However it is important to re-emphasise that many foods of animal origin contain specific substances that are known or suspected to constitute a risk to health. These include saturated fats, the haem products in red meats, and nitroso-compounds in processed meats. The evidence is strong enough to indicate that these foods should be eaten in moderation, but not strong enough to induce a sense of paranoia (WHO 2003).

Food from animals should be produced within systems that are *humane*, in so far as they give proper respect to the health and welfare of the farm animals. Minimum standards for the production of food as a commodity are set by law and defined by regulations. Food from husbandry systems that operate to standards that are demonstrably better than those prescribed by national or international law can be marketed as added-value products.

Efficiency is an obvious measure of the value and competitive viability of any system for the production of food from animals. The most obvious short-term measure of efficiency, as perceived by the producer, is income from the sale of food and other products relative to expenditure on feed, housing, labour, veterinary costs, etc. In the context of planet husbandry, measures of efficiency should include longer term measures such as competitive vs. complementary feed, fossil fuel use, and amortisation of capital resources.

Economics as a measure of the value of food from animals implies more than just cost of food and cost of production. As discussed in Chapter 5, it needs to embrace all the values afforded to resources and benefits to seek a fair balance between the (naturally) selfish pressures from the three stakeholders: producers, retailers and consumers.

The final criterion necessary for the evaluation of food from animals within the context of total agriculture is *ecological soundness.* Food should be produced within systems that are free from pollution (e.g. N and P), minimise *net* greenhouse gas production, preserve soil and water courses, conserve and create habitats for wildlife.

Pathways to better, kinder food

The ultimate goal, best, kindest food, should be excellence as defined by all eight criteria in Box 6.1. This is an admirable aim but unrealistic in the real world, not least because there is an obvious need to compromise between the different criteria (e.g. wholesome, humane, economic, ecologically sound). For practical purposes it makes sense to adopt a scientific approach to analysis of the pathways to excellence within each of the eight criteria, then apply a bottom–up ethical analysis (as discussed in Chapter 5) to achieve the fairest compromise between the differing aims.

The criteria most directly affected by animal husbandry are humanity, efficiency and economics, all of which are governed by the health and welfare of the animals within different systems of feeding, housing and management. The exact path towards a definable, achievable goal of very good, very kind food as defined by the criteria of animal health and welfare within the constraints of economics and affordability, will depend on the point of departure. The right path for an industrial chicken farmer producing millions of eggs per annum for the open market will, of course, differ from that of the African woman with ten hens seeking to provide enough eggs for her family plus a few for sale. This section considers pathways to the production of better, kinder food from animals within the broad categories of milk and dairy products, red meat and poultry, and egg production. For each case the point of departure is defined in terms of the major problems of health and welfare arising as a more or less direct consequence of husbandry practices within intensive and extensive production systems as they operate today. It then explores ways to progress from here. Production problems (for the farmer) associated with non-competitive feeding or breeding – e.g. the animals don't grow fast enough, they get too fat too soon – are not considered except in circumstances where poor performance (e.g. growth rate, feed conversion efficiency) may be diagnostic of health and welfare problems. Problems of disease are restricted to the 'production diseases'. A more comprehensive picture can be obtained from the recent UFAW publication *The Management and Welfare of Farm Animals* (2011). Possible solutions, available now or in the future, are presented for the most part in biological or mechanistic terms, e.g. the roles of biosecurity, vaccination and environmental management in the control of respiratory disease, the roles of housing, management and genetic selection in the control of feather pecking in poultry. In brief, this becomes an analysis and set of recommendations as to what can and should be done. The next steps, namely *how* best they might be done, through implementation of best practice based on current knowledge, improved procedures for quality control, the application of new science, or changes in societal demand will be explored further in subsequent chapters.

Dairy products

Dairy farming in the developed world has moved a very long way from the position where the individual cow was a valued member of the family. In the developed world of industrialised agriculture, herds of more than 100 animals are the norm and massive dairy factories in which cows are permanently confined in herds of over 1,000 animals are becoming increasingly common, not only in the traditional dairy producing regions of Europe and North

America but also in arid zones around the major conurbations of China and the Arab countries. Now and for the foreseeable future, most of the milk and dairy products consumed by most of the world will come from cows in intensive production systems. Moreover, in most of these systems the cows will be housed throughout most or all of their working lives (i.e. from the time of first calving). The comforting promotional image of dairy cows lying at peace in green fields is today an exception. The most successful exception is New Zealand, which has established a major dairy industry and export trade on the production of milk from grass.

In the last 30 years annual lactation yields from the most productive herds of dairy cattle have increased from less than 5,000 to over 10,000 litres. These total lactation yields correspond to peak yields of approximately 25 and 50 litres/day. This has been achieved by parallel processes of genetic selection for increased yield and changes in diet and feeding practice designed to increase nutrient supply. It is important to stress that intensification of dairy production does not lead inevitably to abuse of the principles of good husbandry, whether measured in terms of animal welfare or environmental quality. Indeed some of the worst abuses of welfare (both animal and human) can arise on small family farms from problems that are as old as agriculture, namely failure to provide through poverty, age and illness; in short, failure to cope. Cows in large commercial herds adequately looked after by people and machines are less likely to suffer from catastrophic failures of provision. There is however evidence to show that cows in large commercial units can suffer from systematic failures to provide the quality of husbandry necessary to meet satisfactory standards of welfare as defined by the five freedoms. Some of the most important failures of provision are outlined in Table 6.1 and explained within the following bullet points.

- A dairy cow may both suffer and fail to sustain fitness through hunger, malnutrition or metabolic disease if she is unable to consume or digest sufficient nutrients to support her genetic and physiological potential to produce milk. The risk is greatest in the first weeks of lactation.
- She may suffer chronic discomfort if housing design, especially the design of her lying area, is inappropriate to her size and shape. Problems of poor cubicle design and inadequate bedding may become worse if she loses condition through malnutrition.
- She may suffer pain through lameness or mastitis.
- She may be at greater risk of infectious disease either through increased exposure to infection or increased susceptibility in consequence of metabolic stress.
- She may be bullied or denied proper rest by other cows. She will, almost inevitably, suffer the loss of her calf shortly after birth.

These potential sources of poor health and welfare can be interdependent and additive. For example, the high genetic merit dairy cow, housed in cubicles and fed a diet based on wet grass silage and concentrate in parlour, may suffer both

Table 6.1 Abuses of the 'Five Freedoms' that can arise through systematic failures in the provision of good husbandry for dairy cows

Hunger	Nutrition fails to meet metabolic demands for lactation
Chronic discomfort	Poorly designed cubicles, inadequate bedding
Pain and injury	Claw disorders (sole ulcer, white line disease) Damaged knees and hocks Mastitis
Infectious disease	Mastitis, digital dermatitis
Fear and stress	Rough handling, bullying, separation of calves at birth
Suppressed behaviour	Suppression of grazing, resting, maternal behaviour
Exhaustion	Emaciation, infertility, premature enforced culling

from hunger and chronic discomfort, partly because the quality of feed has been inadequate to meet her nutrient requirements for lactation and she has lost condition, partly because the wet silage has contributed to poor hygiene and predisposed to foot lameness, and partly because genetic selection has created a cow too big for the cubicles.

The Five Freedoms have proved to be a valuable tool for assessing the welfare state of an animal, or a herd of animals, at a moment in time. However for many farm animals, and for dairy cows in particular, some of the most severe welfare problems arise from the long-term consequences of trying, and ultimately failing, to cope with the chronic and exacting physiological and behavioural demands of everyday life. Exhaustion, the final welfare concern listed in Table 6.1, is not identified within the Five Freedoms but is probably the biggest welfare problem of all. It describes the cow broken down in body, and probably in spirit, through a succession of stresses arising from improper nutrition, housing, hygiene and management exacerbated in many cases by inappropriate breeding due to selection for production traits at the expense of fitness. She is likely to be emaciated and infertile. She may be chronically lame. Such cows will be culled because they are no longer productive. Too many cows considered genetically superior in terms of potential milk production are culled far too young (i.e. after three lactations or less). This is not only a powerful indicator of poor welfare for the cows but also a source of considerable economic loss to the farmer since a dairy cow needs to complete at least four lactations to recoup the cost of rearing her as a heifer to the time that she delivers her first calf and starts to generate income.

The three major reasons for culling cows prematurely as no longer fit for purpose are infertility, mastitis and lameness. These are all production diseases: i.e. most of the hazards arising can be directly attributed to the production system. The generic approach to risk analysis and prevention of production diseases was described in Chapter 3 (Figure 3.2). It is necessary to identify the hazards within the system and to measure the consequences for health and welfare (e.g. incidence of infertility, mastitis, lameness). It is then possible to create a risk scenario from measures of the intensity of the hazards, acting singly or in combination, and the severity of the outcomes. This information can be used in a herd health and welfare plan formulated according to HACCP (hazard analysis and critical control points) principles: i.e. identification and analysis of principle hazards and prioritised actions at the most critical control points (Hulebak and Schlosser 2002; Bell *et al.* 2009).

Risk management and husbandry improvements within intensive dairy systems

The greatest hazard for the modern high-yielding dairy cow (typically a Holstein) lies within her own phenotype. Risks within the production system itself may be classified and ranked, in order of increasing intensity, under feeding, housing and management.

Breeding

Near-universal access to artificial insemination has enabled dairy farmers to select semen from bulls proven to be of superior genetic merit measured in terms of the productivity of their progeny. With the globalisation of genetic selection within the dairy industry, more and more dairy farmers are selecting semen from fewer and fewer bulls perceived to be higher and higher up the pyramid of genetic merit. It is also possible to accelerate selection of cows through multiple ovulation with embryo transfer, through cloning, or potentially through laboratory-based genetic modification. These things will be discussed in Chapter 7. Increasing the potential for milk yield necessitates the formulation of diets with high concentrations of metabolizable energy (ME), much higher than that of fresh or conserved grass. The high genetic merit Holstein has been bred to live in a barn not a field.

It is now generally acknowledged by the breeding companies that there has been overemphasis on milk yield (in first lactation) relative to other traits linked to sustained fitness. Table 6.2 (from Pryce *et al.* 1997) describes the phenotypic and genotypic correlations between selection for increasing milk yield, calving interval (a measure of infertility), mastitis and lameness in UK Holstein/Friesian cows. The phenotypic correlation describes the association as it appears on farm, the genotypic correlation is attributable to breeding rather than management and is positive and highly significant in all cases. The negligible phenotypic effects on mastitis and lameness reveal that farmers are just managing to keep things under control through improvements in husbandry designed to compensate for the genetic deterioration in fitness. In the case of infertility, the battle is being lost.

Fortunately for all, dairy farmers and breeders are becoming increasingly aware of the need to modify selection indices to produce a more robust cow (e.g. the Profitable Lifetime Index, PLI: Winters 2007). Whereas ten years ago over 75 per cent of selection pressure would typically be directed towards

Table 6.2 Phenotypic and genotypic correlations between milk yield and three indices of fitness in the dairy cow (from Pryce *et al.* 1998)

	Phenotype	*Genotype*
Calving interval	+0.20	+0.39
Mastitis	–0.01	+0.26
Lameness	+0.04	+0.17

production *per se* (yields of milk protein and fat), modern selection indices give increasing emphasis to traits linked to robustness and leading to improved lifetime performance. Indeed within Holstein UK, fitness traits now contribute just over 50 per cent to the selection index. This is undoubtedly a step in the right direction, though it will take at least five years for its effects to become apparent.

Feeding

Cattle are natural grazers. The rumen has evolved to accommodate a population of microorganisms that steadily ferment digestible fibres (principally cellulose) to supply ME to the host animal principally as volatile fatty acids (VFA). A beef cow grazing moderate pasture may obtain enough ME to support maintenance plus 10–15 litres of milk per day, which is more than enough to support her and her calf. A Holstein dairy cow grazing very good pasture may be able to support a milk yield over 30 litres/day. This falls well short of the 60+litres/day yields obtained from these cows when given *ad libitum* access to mixed rations in barns. In order to meet this very high demand for ME within the constraints of appetite it is necessary to formulate diets of high digestibility. Conventionally, this has meant substituting starch from cereals for cellulose from grasses. This presents two problems: greater dependence on competitive rather than complementary feeds (Chapter 2) and greater risk of rumen acidosis arising from too-rapid fermentation. The most effective solution to both these problems is to make maximum possible use of complementary by-products of crops grown after removal of most of the starch or sugar for human use. Three examples within this wide product range are maize gluten, sugar beet pulp and brewers' grains. I confidently predict that a high proportion of future crop production for energy rather than food use will be directed to the production of fuels for mobile vehicles (e.g. automobiles and tractors) that cannot conveniently be powered by renewable or nuclear electricity. This will generate very large quantities of by-products highly suitable for feeding to ruminants though not to simple-stomached animals like pigs and poultry. Within this context, the capacity of the dairy cow to produce more than she consumes when measured in terms of food for human consumption (Table 2.2) will become ever more valuable.

Feeding-related risks for the dairy cow are greatest in the first weeks of lactation, when she has to accommodate a three to fivefold increase in requirements for ME (Table 2.3) and other nutrients, most critically calcium. This obviously requires major adaptation, both digestive and metabolic, to accommodate the major changes in the quantity and quality of feed. If the cow cannot take in enough ME to support the demands of lactation she will lose condition and be at greater risk of infertility. If her metabolic hunger drives her to consume more highly digestible feed than her rumen can process, she is likely to experience recurrent bouts of rumen acidosis. This will cause her distress in the form of pain and malaise, and impair her appetite, leading to loss of milk production and body condition. Fortunately the biological causes of these problems are relatively well understood by better informed dairy farmers and veterinarians. This has led to significant improvements in 'transition' feeding and management during the critical months before and after calving. It is fair to conclude that these problems can be and are being satisfactorily managed in the best intensively managed, high-yielding dairy herds. While many high-yielding dairy cows may currently be at risk from improper feeding practices, it is fair to conclude that these problems are manageable. Thus I predict that the contribution of improved feeding to the production of better, kinder milk from dairy cows will be achieved less through the acquisition of new scientific knowledge than through the wider implementation of current best practice.

Housing and environment

Dairy cow housing is designed primarily for the convenience of the farmer, to facilitate feeding, milking and handling. However, buildings with roofs and concrete floors are expensive, which means that the space allowance per animal has to be restricted. Dairy cows, given the chance, elect to spend half their time lying down (12/24 hours). This means they need a comfortable bed. Selection for milk yield in the Holstein cow has created bigger and bonier animals that are more susceptible to injury when lying on hard or abrasive surfaces and more likely to have difficulty in the acts of standing up and lying down. The two main options for lying areas are individual cubicles (free stalls) or bedded yards. Traditionally small herds of cows in mixed farms would lie on a deep bed of straw, which is comfortable and warm though a potential hazard for mastitis. Modern large specialist dairy farms are unlikely to produce enough straw on farm so usually opt for cubicles, which are more expensive to install but reduce bedding costs. Individual cubicles offer cows more security than densely stocked yards but they can restrict freedom of movement (standing up and lying down) and in many cases the lying area is profoundly uncomfortable. The principal hazards in dairy cow houses are hard and injurious floor surfaces creating the risk of chronic discomfort, painful injuries to feet and legs, and infections arising through poor hygiene, e.g. environmental mastitis and digital dermatitis. Badly designed houses that restrict freedom of movement, especially around feeding and drinking areas, can also create problems of aggression.

Historically, most cows were tethered in individual stalls throughout the winter months and milked where they stood. In urban dairies, they were tethered for most of their working lives. On many small, traditional dairy farms the practice of tethering cows over winter continues today. Scientific and popular opinion within Europe considers this practice an unacceptable restriction on normal behaviour. On the other hand, 'progress' from the time when small herds of small cows were fed dry hay as forage in tie stalls to the present, where large herds of large cows are fed wet silage in cubicles, has significantly increased the risk of lameness: currently the top priority for action to address welfare problems in the dairy industry.

This begs the question: 'Is it possible to design and build a house that fully meets the needs of the dairy cow for comfort, security, hygiene and reasonably free expression of behaviour?' This has particular significance for the design of 'zero-grazing' units in which cows are housed throughout the year, or for the duration of their lactation, with two months holiday at pasture, or in yards, during their dry period. One of the most popular criticisms of zero-grazing systems is that they deny cows the most natural and time-consuming practice of grazing at pasture. This is self-evidently true. However the scientist should examine in more detail the ways by which pasture may contribute to good welfare.

- Fresh grass is nutritious and consistent with healthy digestion, although not nutritious enough to meet the metabolic requirements of high yielders. Moreover, on many farms cows spend much of their time outdoors on 'sacrifice' pastures that provide little in the way of nutrition; merely somewhere to rest – and not muck up the yard.
- Reasonably dry, sheltered pastures provide excellent physical and thermal comfort in most circumstances. However, lack of shade from the sun can exacerbate heat stress during hot periods. Wet muddy pastures increase the risk of discomfort, foot injuries and udder infections.
- Cows at pasture almost invariably have much more space than indoors, which gives added security and freedom to indulge in, or avoid, social contact. However, cows that have to walk long distances twice daily from pasture to parlour are at increased risk of lameness from foot injuries.

From the cow's point of view, therefore, the greatest attractions of pasture are comfort, security, socialising and (to give them the benefit of the doubt) fresh air and sunshine. These needs cannot be met in a building where cows are housed permanently under a roof, rest in cubicles and walk on concrete. They can however be met in a zero-grazing unit that provides ample, comfortable resting space, ideally a broad expanse of deep sand, and free access to an outside loafing area. I have seen many such units on many continents and the best have been very satisfactory.

A fashionable argument in favour of the highly intensive, fully roofed, fan-ventilated cubicle house is that it can be more environmentally friendly, through reduced pollution from contaminated rain water, improved recycling of wastes using an anaerobic digester and possible capture of methane emissions. This could be the best solution in terms of the criterion of ecological soundness, provided the installation costs are properly taken into account, but I do not rate it highly on grounds of humanity.

Management

The most serious current problems attributable to management are:

- lameness and environmental mastitis attributable to poor hygiene;
- lameness attributable to inadequate foot care;
- inconsistent stockmanship;
- unwanted male 'bobby' calves.

Modern high-yielding dairy cows consuming very large quantities of wet silage produce prodigious quantities of very wet faeces. When such cows are confined in barns or cubicle houses this presents major problems of manure removal and disposal. Correct manure disposal is obviously critical in the matter of pollution control. Diligent removal of manure from the floor and especially the bedding area is essential to reduce, though not eliminate, the risks of lameness and environmental (coliform) mastitis. Bedding can never be completely hygienic, although sand has been shown to constitute a lower risk than other substrates such as straw and wood shavings.

Foot lameness in dairy cows has two main causes, claw lesions such as sole ulcer and white line disease, and skin infections such as digital dermatitis. The risks of both are high in cubicle houses with concrete floors and poor hygiene. However there is good recent evidence that the greatest risks for total and severe lameness arise from inadequate attention to foot care, involving regular foot inspection with trimming as appropriate to the individual cow, regular foot bathing for the entire herd in a suitable disinfectant, and especially diagnosis and treatment of animals as soon as they show departure from normal locomotion. This is a particularly urgent example of a serious general problem for dairy cows and their farmers. In large dairy units operating to tight profit margins, farmers and herdsmen do not have sufficient time to look at their cows properly, let alone given them sufficient individual care. Two other factors contribute to this problem. Overworked farmers become habituated to things they see every day. Typically their estimate of lameness prevalence based on observation of locomotion will be four times lower than that of an independent trained observer. The limping cow is seen as normal. The other main hazard is that cows in large herds tend to be milked by a succession of contract milkers who see little more of the cows than their back ends in the milking parlour. This makes it almost impossible to carry out an effective policy for inspection of individual cows for evidence of problems of health and welfare.

It is probably unrealistic to expect a significant reversal in the trend towards reduced labour input in large commercial dairy herds. The biggest problem is that nearly all the labour time is taken up in routine chores such as milking, dispensing feed and manure disposal. This problem calls for alternative solutions. One is through the application of new technology. The development of the robot milker has reduced the chore of milking and there is no significant evidence to indicate that it impairs cow welfare. Moreover the milking station that the cow enters at her own volition creates opportunities for monitoring a wide range of health and welfare indicators, including reproductive status, incipient mastitis and lameness, and nutritional disorders. Another route to improved husbandry through improved attention to management is through the implementation and promotion of quality assurance schemes, backed up by guaranteed robust protocols for quality control. These are discussed in Chapter 9.

The final big problem for the specialist dairy industry is the problem of male 'bobby' calves, unwanted by the beef industry by virtue of their singularly unmeaty conformation. These calves, having little or no value, are likely to be killed shortly after birth or, which may be worse, condemned to life in intensive units for the production of white veal (see below). The three most constructive solutions to these problems are:

- increase selection for a more dual-purpose cow, whose calves are of value both for milk and meat production;
- prolong lifetime performance in the lactating cow through selection and management;
- increase use of sexed semen to achieve the optimal sex ratio for the dairy herd.

The dual-purpose cow (with a lower lactation yield) is likely to receive more favour within pastoral systems than within the high-input/high-output zero-grazing units. Increased lifetime performance (e.g. from <3 to >5 lactations) increases the opportunities for dairy farmer to use semen from beef bulls, effectively reducing the number of dairy-type males from one in three to one in five. The third and conceptually most satisfactory solution is through the use of sexed semen. At present the cost, fertility and genetic merit of the available sexed semen are less than ideal but the approach shows great promise.

Pastoral systems for dairy cows

Traditionally dairy cows subsisted on grass and other complementary feeds, most of which they foraged for themselves. The dominance of the specialist high-yielding Holstein, bred to live in a barn, has been driven by the inexorable economic forces that govern the supply and demand of commodities. If pastoral systems for dairy cows are to succeed, they have to compete either as commodity producers, or as specialist suppliers of added-value products.

There are some areas of the world where grass is still the most economic feed for dairy cows. The massive development of milk from grass in New Zealand, mainly as a source of butter and cheese for export, is proof that milk from grass can compete, without subsidy, in an international free market. Despite advertising claims that this system is 'free range', it is less idyllic than it may appear. Herds in excess of 500 animals are being kept on intensively managed monocultures of grasses and walked very long distances to and from the milking parlour. These large herds are creating new problems for the cows, especially increased prevalence of lameness, and for the environment, through destruction of mixed, sustainable ecosystems (Mairi 2007). Pasture-based dairy production continues to be a major contributor to mixed farming in traditional cow country like the south-west of England where I live. It is efficient in that most of the feed is grown on farm and complementary rather than competitive. Most of the male calves are reared for beef. It is compatible with the ecology and aesthetics of the living environment. As I have written elsewhere:

> Dairy cows at pasture have made an invaluable contribution to the sensitive, sustainable and downright beautiful stewardship of the British countryside. The cows are moved through small fields with high hedges that are a Mecca for wildlife and the lanes between the fields are an explosion of wild flowers. When the cows come in for milking, they bring in the muck but the muck attracts the insects and the insects attract the swallows and martins that nest in the eaves and the swifts that soar, and sleep, in the sky. If this is what the urban majority wish to preserve for our own peace of mind, we must be prepared to acknowledge that it is worth more than the price we pay for milk, either directly or via taxation through redirection of agricultural subsidies.
>
> (Webster 2011b)

The high genetic merit Holstein bred for a short and exhausting career in a barn is not the ideal animal for pastoral systems because, as explained earlier, its metabolic capacity to convert nutrients to milk exceeds its digestive capacity to acquire nutrients from grass. This has, eventually, been recognised by this sector of the dairy industry, who are now developing a more robust, beefier animal whose slightly lower milk yield is compensated for by a reduced demand for purchased concentrate feed and increased income from the sale of male calves for beef.

The high-welfare, organic image of cows in green fields has obvious appeal to consumers for whom the ethics of the production system is a 'non-use' measure of the value of dairy products. Pathways to the promotion and surveillance of pastoral 'free range' systems of dairy production will be explored in Chapter 9.

Traditional and village systems

In much of the less developed world, lactating ruminants, e.g. cows (*Bos Taurus* and *Bos indicus*), buffaloes, yak, continue to be indispensible contributors to the welfare and income of pastoral and village communities. They produce highly nutritious food in circumstances where the demand is great and the alternatives are few. The animals are well adapted to challenging environments but, in consequence, milk yields are low. There is undoubtedly scope for some cross-breeding between indigenous and 'improved' breeds but this must be done with care and a proper understanding of local circumstances in order to create a phenotype appropriate to the local climate and feed resources. For some years there was a UK charity that urged us to 'send a Holstein to Africa'. That was more than just silly. It was unfair to the cows.

The most fruitful pathways to the production of better, kinder dairy products within traditional pastoral and village systems involve the implementation of existing knowledge and fair allocation of resources, especially in relation to the control of infectious and parasitic diseases. One of the greatest problems for the herders is that they cannot afford veterinary care. Investments in veterinary care by governmental and non-governmental organisations (e.g. charities) can create real wealth by removing major constraints to the capacity of animals to convert local resources into food. The other big contribution has to be through education, sensitive to and complementary to the substantial knowledge of the indigenous people.

Beef

In areas of the world such as the prairies and pampas of North and South America, beef production is based on a population of suckler cows on range that produce and rear one calf per year. After weaning at six months or more, calves will be removed and finished for beef either at pasture or, more likely, in feedlots where they are fed large quantities of cereals, including ensiled whole-crop corn (maize). Although the consumption of competitive cereals by beef cattle in feedlots is high, the overall proportion of competitive feeds consumed is only one-third (Table 2.2) because half the feed is consumed by the mother cows (Table 2.1). In Europe, more than half the beef comes from calves born to dairy cows inseminated with semen from beef bulls (e.g. Charolais). These calves are taken from their mother shortly after birth and artificially reared and weaned on to solid feeds as soon as possible because income from sale of milk to the public is greater than income from milk fed to calves. Subsequently they will be reared for beef in a variety of systems ranging in intensity from calves permanently housed in barns, fed predominantly on cereals and killed at 12 months, to animals finished at pasture at 30 months or more.

Cows and calves on open range are usually well able to look after themselves. When there are welfare problems these tend to result from climatic disasters (droughts, storms) rather than systematic failures of husbandry (feeding, housing and management). When the calves enter feedlots and we assume control of their destiny then problems arise. Artificially reared calves from the dairy herd are at risk from systematic failures of husbandry from the moment of birth.

Table 6.3 highlights the most important husbandry-induced risks to the health and welfare problems for the slaughter generation of growing beef cattle. The relative importance of the different risks is indicated by asterisks. Only the more severe problems will be discussed in any detail. Suckler calves taken off range

Table 6.3 Health and welfare risks for finishing beef cattle in different systems.

	Feedlots	*Housed systems*		*Pasture*
		High concentrate, straw bedding	*High forage, concrete slats*	
Feeding	Acidosis*** Laminitis**	Acidosis*		
Comfort	Thermal stress*		Floors**	Thermal stress*
Pain and injury	Branding* Transport*	Foot lameness*	Parasitism*	
Disease	Pneumonia***	Diarrhoea** Pneumonia**	Diarrhoea** Pneumonia**	Diarrhoea* Pneumonia*
Behaviour	Bullying*	Bullying*	Restricted space** Bullying**	

Asterisks indicate the relative importance of the risks.

and transported to large, open feedlots (Figure 1.2) are presented at once with a number of severe stresses, often lumped together as 'shipping fever'. These stresses include the handling and transport of animals previously unaccustomed to human contact, mixing of immunologically naïve animals with those carrying respiratory infections, and abrupt changes in diet from grazed pasture to cereal-based concentrates. The most severe and immediate problem can be an outbreak of fibrinous pneumonia of bacterial origin with mortality and morbidity rates ranging from 5 to 20 per cent and 20 to 60 per cent respectively (Schneider *et al.* 2009). Growth rates are usually impaired in recovered animals. Approaches to the control of pneumonia on feedlots include preconditioning, including the use of vaccines, reducing transport stresses and better control of the changes in diet. Future possibilities for control include the genetic selection of resistant animals. These are discussed in Chapter 7.

The high-starch rations for beef cattle finished on feedlots carry a high risk of inducing chronic disorders of digestion arising from unstable rumen fermentation leading to acidosis. The direct consequences of acidosis are discomfort and loss of appetite. Common sequelae of acidosis include chronic disruption of the rumen epithelium (parakeratosis) and liver abscesses: morbidities can exceed 50 per cent (Corbière *et al.* 2008). A less common but more severe consequence is laminitis, a condition arising from inflammation and capillary damage within the horny laminae of the hooves. The pain from acute laminitis is very severe; the pain from chronic laminitis remains severe enough to compromise feed intake and growth rates to an extent sufficient to cause significant economic losses. The animal health and welfare problems listed in Table 6.3 can be severe but the risks can be reduced through better management based on the application of existing knowledge.

The most important risks for artificially reared calves born to dairy cows are infections caused by endemic viruses leading to diarrhoea and pneumonia. The incidence of diarrhoea is greatest in the first six weeks of life, before weaning. When the probability of exposure to the endemic viruses is very high, the most effective approach to risk management is through attention to hygiene and feeding, especially the feeding of liquid milk replacer diets, in order to avoid indigestion, which predisposes the gastrointestinal tract to infection. Vaccines against the more common viruses are available but do not guarantee protection. Unweaned calves that suffer from diarrhoea are at greater risk of pneumonia. The risk of pneumonia is also high in all intensive beef rearing systems when animals are housed throughout the rearing period. The long rearing period (relative to pigs and poultry) means that these units seldom operate an 'all in–all out' policy. In consequence more or less immunologically naïve calves are constantly exposed to infection from sick and recovered carriers. The most effective approach to risk management is through preconditioning, as with animals destined for feedlots, based on strategies involving concern for the provenance of animals, vaccination as appropriate to need and reduction of stresses associated with movement and adaptation to new diets.

When calves from the dairy herd are reared for slaughter at 12–14 months on high-cereal diets, their housing tends to include a bedding area of straw in deep litter. This is practicable partly because cereals and straw go together and partly because dry feeds generate relatively dry faeces. Such calves are physically comfortable but at relatively high risk of acidosis and the other digestive disorders described above. Unless the deep litter of straw bedding is constantly topped up and kept dry, the calves are at risk of lameness associated with infected, damaged feet, the infections arising from within the bedding, the feet probably damaged during episodes of laminitis. Units that rear calves on high-forage diets, predominantly grass and maize silage, are less likely to have straw on farm. Moreover the calves produce more urine and wetter faeces. In these circumstances it is common to house the animals permanently on slatted concrete floors, which are obviously uncomfortable and increase the risk of injury. They are also stocked at high density, partly because such buildings are expensive and partly to increase the rate at which manure is tramped through the slats, so maintaining a modicum of cleanliness. This reduces the opportunity for the cattle to lie down without risk, or fear, of being trampled on by others. The high stocking density also increases the risk of respiratory disease.

Behaviour-related problems for large beef animals stocked in large numbers and at high density arise partly from deliberate acts and partly by accident (e.g. animals being stepped on while lying down). Most of the male calves from the dairy herds of Europe that are permanently housed and reared intensively for beef are kept as entire bulls because they grow better on high-energy feeds. These animals can display a great deal of sexual behaviour (called 'bullying' for short in Table 6.3), which mostly takes the forms of animals riding one another in simulation of mating behaviour. This can lead to disturbance, injuries and loss of condition in some individuals who appear to be singled out as favoured recipients. In some European bull fattening units, wooden boards or even electrified wires are spread over the units just above animal height to prevent this natural but antisocial behaviour. As with individual stalls for pregnant sows, it is an engineer's solution that resolves the problem, as seen by the producer, who caused it in the first place by confining the animals in an environment entirely unsuitable to their behavioural needs.

The fourth category of systems listed in Table 6.3 is that which involves rearing animals out of doors on pasture for as long as possible, housing them in yards on a diet of conserved silage, and feeding the minimum amount of concentrate necessary to ensure that they are in satisfactory condition at slaughter. This is obviously the most natural system, although it is necessary to castrate the male calves, partly to prevent random mating and partly to assist fattening on relatively low-energy diets. Unsurprisingly, health and welfare problems are few. Table 6.3 identifies calf diarrhoea and pneumonia as possible risks for artificially reared calves from the dairy herd. These are likely to occur in the first months of life before the calves are put out to pasture. The risk is less than for calves housed throughout the rearing period because there is a much lower probability of vertical transmission of infection from older carrier animals. Finishing beef cattle at pasture is traditional in high rainfall beef rearing areas such as Ireland where grass grows well and cereals don't. In recent years beef from animals finished at pasture has grown in popularity as an added-value product for the discerning consumer. Claims for added value include local provenance (e.g. Scotch beef), organic, high welfare (e.g. Freedom Food), improved taste and (with less assurance) reduced health risk.

Beef production, especially the mass production of beef from cattle on feedlots, has drawn criticism on multiple grounds: pollution, environmental consequences of mass demand for concentrate feeds such as soya beans, and human health risks associated with high consumption of red meats. The environmental and health risks are very real and, as discussed in Part I, the future for animals in sustainable agriculture will, sooner or later, have to include less beef, for their sake and ours. Within the specific context of this chapter, 'Better, kinder food', I confess I cannot offer any radical solutions to health and welfare problems within large intensive production systems other than improved attention to the management of known risks. The greatest promise for all parties within the ethical matrix (Table 5.1) lies in the increased acceptance by society that beef production is best suited to well managed areas of extensive grassland and parkland (grass and trees) where outputs are low but the system is not only sustainable but positively beneficial to the environment through carbon sequestration, management of water courses, amenity, encouragement of wildlife and many other etceteras. This inevitably implies that beef will be harder to come by and more expensive, which is healthy. However, to return to my main theme, sustainable beef production should ideally be viewed as one vital component of total agriculture within planet husbandry and rewarded in such a way as to make it competitive with intensive commodity production of beef in feedlots.

Veal

Since the second agricultural revolution, i.e. the advent of factory farming, the conventional form of veal production has been the production of white veal using male calves from the dairy herd and rearing them to an age of four to eight months entirely or almost entirely on a liquid diet consisting largely of dried skim milk products with added animal fats. This industry evolved to harvest by-products from the dairy industry, namely fat-free milk and male dairy-type calves, to produce white veal, a tender meat, lacking the cherry-red colour and taste of mature beef. The whiteness of the meat is achieved by restricting the iron concentration of the ration to prevent the synthesis of myoglobin that gives the cherry-red colour to muscle. In the process it also reduces the synthesis of haemoglobin in red blood cells to a degree where many calves are clinically anaemic. Furthermore the system legal in Europe until 2007 and still legal in many parts of the USA constituted a systematic abuse of all five freedoms.

- The total absence of solid feed containing digestible fibre prevents the normal development of the rumen, causing abnormal stereotypic behaviour (e.g. tongue rolling), frequent indigestion from hair balls, and a high predisposition to endemic infections leading to diarrhoea, pneumonia and death.
- The abnormally high consumption of milk predisposes to abomasal ulcers.
- Confinement of calves in individual crates denies social behaviour, and makes it impossible for calves to turn round, groom themselves, or adopt normal lying positions for rest and sleep.
- Confinement in darkened houses without access to the sights and sounds of farm activities makes calves especially vulnerable to fear and stress.
- In many intensive units the routine administration of antibiotics to veal calves has been 100 per cent, on the assumption that without constant antibiotic cover, morbidity and mortality from intestinal and respiratory infections would be unacceptably high.

Intensive white veal production is actually a corruption of a practice that predates the first agricultural revolution. When the peasant family's cow gave birth to a male calf, the cow could support her calf at first on milk produced surplus to family needs. It was however neither prudent to feed the calf on solid feed that could be used more productively by its mother, nor to rear the calf to an age and weight when it could no longer be consumed at one large sitting. Thus the fatted calf was reserved for big and special occasions, like the return of the Prodigal Son. The meat would have been white because cow's milk is naturally deficient in iron, not a problem for calves allowed to run with their dams at pasture.

The first stirrings of revolt against this profoundly unnatural and abusive system of meat production came with the publication of Ruth Harrison's seminal book *Animal Machines* (1965), closely followed by the report of the

Brambell Committee of Inquiry into the welfare of animals kept under intensive husbandry systems (1965). The two worst abuses, namely denial of feed containing digestible fibre and individual penning after the age of eight weeks, were prohibited in the European Community from 2007. Life is now a little less dismal than it was for calves condemned to white veal production, but in terms of digestive disorders, predisposition to infectious disease and abnormalities of behaviour it is still, in my opinion, as far from the concept of good kind food as any other legal system of animal production.

In recent years there have been some small-scale efforts to develop higher welfare systems for the production of pink or rosé veal. The primary aims are to provide enough digestible fibre to promote reasonably normal digestion, enough iron to avoid anaemia and enough space and environmental enrichment to give the calves a life worth living. The interpretation of these aims varies greatly, from minimalist with regard to feed and space to near ideal involving the rearing of calves at foot with their mothers at pasture to slaughter at about six months of age, the time of normal weaning. If rosé veal is to be marketed as high welfare, this needs to be guaranteed by an independent procedure for surveillance and quality control, such as Freedom Foods (see Chapter 9).

The deficiencies of white veal production, even within the new European legislation, are intrinsic to the system and not likely to change unless in response to public demand. White veal, unlike eggs from caged birds, but like *foie gras,* comes from an abusive system designed not to provide cheap food for the masses but expensive food for the relatively affluent. The so-called 'discerning' individuals who elect to buy white veal clearly do not include welfare of animals within their parameters of quality and added value. To be fair, this may be because many of them simply haven't thought about it. The solution to this must lie in better communication of the general awfulness of the system. This should not call for hyperbole and sensationalism. The facts of the case are bad enough.

Pigmeat

Pig production in the developed world has become one of the most industrialised and engineered forms of animal farming. Populations of several hundred breeding sows and their several thousand offspring are kept in large controlled environment buildings. To minimise the risk of disease entry the units are surrounded behind security fences. Workers shower in and shower out and change into clothing to be worn only inside the unit. In effect, the whole operation is run like a maximum-security hospital. Since all the feed (mostly cereals) is processed and dispensed by machine, there need be no direct connection between the produce of the meat factory and the produce of the land. The only direct need for contiguous land is somewhere to spread the slurry without incurring formal penalties for pollution or the wrath of neighbours.

Within the most successful units the efficiency of conversion of animal feed to animal product, pork, ham and bacon, is close to the theoretical maximum. The main contributors to increased efficiency have been:

- Genetic selection for rapid, lean growth in the slaughter generation, which has inevitably led to a large increase in mature weight in the breeding sows (cf. Table 2.1).
- Maximising output of live piglets per sow per year (>20) through control of neonatal mortality, early weaning (at three to four weeks), and selection for fecundity.
- Control of infectious disease through strict attention to biosecurity: foundation and maintenance of specific pathogen-free herds, vaccination against specific infections, and (in many but not all cases) copious administration of antibiotics.
- Maximising feed conversion efficiency in the slaughter generation partly through control of the thermal and physical environment (warm and crowded) to minimise metabolic heat production but mainly through the control of low-grade infections.

When the terms of reference for analysis of these enterprises are defined simply by output of meat relative to input of feed, it is beyond dispute that they are much more efficient than more traditional systems in which small groups of sows were kept not in isolation units but exposed to a 'normal' challenge with endemic pathogens, reared their piglets to about eight weeks of age before weaning and produced maybe 12–16 piglets per year. Given a broader audit, the advantages become less obvious. Capital costs for large buildings containing a lot of robust expensive hardware (e.g. farrowing crates) are high, as are fuel costs for machinery and the maintenance of controlled environments. At a recent conference on animal welfare, a researcher with severe tunnel vision (who need not be named) described a more humanely engineered farrowing crate that permitted sows more freedom around the time of parturition. He illustrated this with a table to show that, although it was clearly superior to the conventional farrowing crate in intensive units, it was unlikely to be adopted because it was considerably more expensive. He did not comment on a third column in the table that clearly showed that the cheapest of all options in his study was to allow sows to farrow outdoors in simple arks.

The main health and welfare problems for pigs in intensive units are well documented and reasonably well understood. The major hazards, welfare consequences and measures used to assess welfare consequences are summarised in Table 6.4. I shall not discuss these in detail but consider some more or less radical solutions to the most severe systemic problems (i.e. those inherent to the system).

Farrowing accommodation

The conventional farrowing crate that denies the sow any expression of maternal behaviour, indeed the opportunity to do little more than stand up and lie down, is intended to minimise the risk of piglets being crushed when the sow collapses into a lying position. This is a real risk for piglets but only

Table 6.4 Husbandry hazards and welfare consequences for sows and growing pigs in intensive systems

Husbandry	*Hazard*	*Consequence*	*Measure*
Sows			
Feeding	High energy feeds	Frustration	Bar chewing, aggression
Housing	Hard floors	Injuries, arthritis, pain	Skin lesions, lameness
	Confinement	Osteoporosis	Discomfort, lameness
	Cold floors	Cold stress	Weight loss
	No foraging substrate	Frustration	Aggression
	Farrowing crates	Frustration	Mostly denied
Hygiene	Dirty floors	Urinary infection	Pain, fever
	No vaccination	PRRS	Piglet mortality
Management	Mixing	Aggression	Fighting. Injury
Breeding	Selection for size	impaired maternal behaviour	Piglet mortality
	Selection for fecundity	Piglet mortality	
Growing pigs			
Feeding	Early weaning	Enteritis	Diarrhoea
Housing	Hard floors	Discomfort, injuries	Skin lesions, bursitis
	Poor ventilation	Respiratory infections	Coughing, fever mortality
	No foraging substrate	Frustration	Aggression, tail biting
Hygiene	Low biosecurity	Multiple infections	Mortality, diarrhoea, ill thrift
	No vaccination	PRRS, PMWS	Mortality, ill thrift
Management	Overstocking	Infection, aggression	Fever, fighting, ill thrift
	Mutilations	Pain	Vocalisation
Breeding	Selection for leanness	Stress susceptibility	Sudden death

during the first 48 hours of life. Several comparisons of piglet mortality to weaning show little difference between intensive units with farrowing crates and outdoor units with simple arks. These comparisons need to be treated with caution. Part of the explanation is that sows bred to live outdoors are smaller and better mothers. For the large clumsy sows bred to maximise productivity in intensive units, farrowing crates do save lives but these units should be considered only as maternity units. On grounds of humanity and economics it can make more sense to move sow and litter to a cheaper nursery suite before the piglets are one week of age. This should reduce the requirement for farrowing accommodation by a factor of at least three, and possibly justify a more sympathetic design. There is however strong evidence to show that where the environment (rainfall and soil type) is suitable, the breeding is appropriate and the stockmanship is good, not only will the sows and litters thrive in simple outdoor accommodation but the farms can be economically competitive. In simple terms, it is quite possible to practise good husbandry and survive on a proper pig farm. The answer to those who say that they are compelled to operate an intensive confinement unit because the land is too wet and/or prone to pollution is that you don't own a pig farm.

Early weaning

When the main aim is to maximise output of piglets per sow per year it is logical to wean the piglets and rebreed the sow as early as possible. The limits of this policy are 18–20 days to weaning, plus 7–9 days before return to oestrus, plus 114 days pregnant, giving an inter-farrowing interval of about 140 days. Given a target of 12 piglets reared per litter this can, in theory yield in excess of 30 piglets per year. In practice, few units achieve better than 25 and the average is closer to 22. When animals (pigs, chickens, dairy cows) are driven to their limits, things break down. Weaning piglets at three weeks of age or less is profoundly unphysiological. Their digestive tract is immature and likely to suffer damage when fed a diet of cereals and vegetable proteins. Their immune status is at its lowest and they are extremely sensitive to cold unless they can huddle together in a warm bed of (e.g.) deep straw. To combat these self-imposed hazards, the piglets are given expensive processed feeds and confined in sterile, barren cages in heated buildings ('flat-deck' accommodation). Even then the risks of enteric disease associated with common bacteria are high, and entry of virulent strains of organisms can be catastrophic, e.g. PMWS, the 'piglet multiple wasting syndrome'. Vaccination can provide some control (at a cost) but too many pig units are dependent on routine administration of antibiotics to reduce the economic cost of opportunist bacterial infections in animals routinely damaged by the system itself. This is a classic illustration of the premise that the optimal strategy for the herd, measured in terms of gross profit margin, will often conflict with the optimal strategy to ensure the welfare of individuals within the herd.

In theory, the effective implementation of a ban on the routine prophylactic administration of antibiotics to growing pigs would enforce better standards of husbandry on the pig industry (after a transition period during which more piglets would die). However, implementation is easier said than done, not least because veterinarians, acting with varying degrees of integrity, can prescribe large quantities of antibiotics, not prophylactically but nominally for treatment or metaphylactic purposes (i.e. medication of those at risk in the known presence of disease carriers).

Environmental enrichment

The pig's nose is a wonderful tool; at least 100 times more sensitive to smell than ours, yet tough enough to act as a plough or even, in my own experience, to undermine the wall of an ancient pig sty. Pigs are highly motivated to sniff around and root in the ground. Food is their obvious goal but there is more to their motive than hunger. Foraging is highly satisfactory so denial of foraging behaviour is profoundly frustrating. For sows in groups this frustration may give rise to fighting and injury. Sows isolated in pregnancy stalls show a high prevalence of stereotypic bar-chewing. Piglets reared in barren pens without bedding or 'foraging substrate' are more prone to tail biting. Bitten tails can become infected leading to painful abscesses. Sometimes the sepsis can enter the spinal cord and cause death. Tail biting in pigs, like feather pecking in hens, is not primarily an aggressive act, the tail merely acting as just about the only thing worth investigating in an otherwise barren environment.

In the EU, it is now compulsory to provide growing pigs with some form of environmental enrichment. In many cases, however, this enrichment may satisfy the legislation but not the pigs. Pigs become quickly bored with 'toys' such as tennis balls, chains and rubber tires because they do not offer any reward. Rewards do not have to be immediate and generous; sows and piglets will root for hours for very little so long as they have some hope of success. Ethologists will recognise similarities in the behaviour of sows foraging for worms in the mud and punters working the slot machines in Vegas. Hope is all.

The decision to allow pigs to run and forage outdoors raises the question 'is it acceptable to fit them with nose rings to prevent them from ripping up the pasture?' There is no doubt that sows, given time, will reduce any pasture to the status of a badly ploughed field. On balance they probably do more harm than good to the fertility of the soil, not least because of their appetite for earthworms. However, they do clear the weeds. Pig farmers on the well-draining soils of the English South Downs use their herds of unringed sows as a highly successful crop rotation between years of cereals. Sows allowed free range, e.g. in apple orchards, are usually equipped with nose rings. This deters their natural urge to root up the ground. However they are still able to range, seek and consume anything they can find at ground level, from ants to apples. This has to be better than life on concrete. Weaned, well-fed young pigs running free at pasture or in orchards seldom need nose rings.

Futures for pigs

It is clear that the conventional, confined, intensive production system for pigs is incompatible with the needs of the animals for good health, comfort and freedom from injury, oral satisfaction and appropriate behaviour. It is a system that has been designed by engineers with little or no consideration, *a priori*, of the pigs' own criteria for suitability. Tinkering with the system, e.g. redesigning farrowing crates for sows, docking the tails of young piglets, does no more than plug some of the leaks in a system that is fundamentally flawed in concept. Within much of Europe and North America, the economic advantages of intensive systems relative to alternative systems in which the sows and piglets (outdoor breeding) or all pigs (outdoor rearing) are kept outside are small and disappear when the outdoor systems can be marketed at a premium price. It is undeniable however that outdoor systems, operating only on land suitable for such systems, cannot meet our current demand for pig meat. One healthy solution to this problem is, of course, to eat less. It is unrealistic to expect this goal to be achieved by legislation but there are good reasons to hope for slow but steady progress to better husbandry in pig production through increasing public demand for pig meat from freer range alternatives.

On a world basis, it is certain that consumption of pig meat will increase. Even in those countries that have cultural and religious objections to the practice it is probable that increasing secularisation will bring the bacon into more homes. In China, where more pigs are eaten than anywhere else, pig farming practices range from the traditional peasant family unit to the largest of peri-urban total confinement industrialised production systems. I have visited large production units, livestock markets and abattoirs in China and found them to be essentially the same as those operating in the West 30 years ago before the upsurge of public concern about production standards, antibiotic usage and animal welfare: i.e. efficient but lacking compassion. My greatest concern about these systems is not that they involve deliberate acts of cruelty but that they have been designed for the efficient processing of units of livestock into meat, with little sympathy for or attempt to understand the needs of the sentient animals whose lives we switch on, drive and switch off in accordance with our own wishes. I have, in China, seen evidence of acts that would appear appallingly cruel to anyone with any feel for or understanding of animal sentience, such as the use of gaff hooks to drag pigs out of travelling cages like a sack of grain. In fact, this is even more stupid than cruel, since pigs, unlike sacks of grain, vigorously resist the process. My colleagues at Bristol supervised the installation of a handling system that allowed pigs to move naturally and spontaneously at all stages to the point of slaughter (as is standard in Europe and North America), and the owners of the abattoir were delighted because more pigs could be moved faster and with less labour. I repeat, the problem was not so much cruelty but ignorance: a total failure to understand what it is to be a pig.

While in China I also met a number of people, especially young people at agricultural and veterinary institutes, who were highly motivated by the concept of farm animal welfare and keen to learn. I was also particularly impressed by the attitude of many senior members of these institutes who were extremely open-minded and receptive to alternative arguments and ideas. The industrialisation of pig production in China and South-east Asia is running about 30 years behind the West, not only in terms of industrial practices but also in terms of the proportion of pigs involved in industrial production. Thus, when designing new industrial units, as they surely will, they have the opportunity to avoid some of our worst mistakes in relation to housing and management practices. From the pigs' point of view, it will be far from ideal, but at least the worst abuses can be avoided.

Poultry meat

Broiler chicken production presents the most extreme manifestation of the impact of the second agricultural revolution, the industrialisation of agriculture, both on production methods and on the cost and availability of the end product. Sixty years ago chicken meat was a luxury, now it is the cheapest of all meats, currently in my local supermarket, about half the price/kg of cat food. The main reason why it has been possible to bring about such an enormous change is very simple and was illustrated in Table 2.1, namely the high prolificacy of the breeding hen, which means that chicken feed can be apportioned 96 per cent to the slaughter generation, 4 per cent to the breeding generation. This makes it economically expedient for chicken breeders to focus exclusively on traits relating to the rapid growth of lean meat at high feed conversion efficiency, without regard for traits relating to sustained fitness in adult life. The result of this breeding strategy has been genetic strains of birds that can achieve slaughter weights of 2–3kg at 38–42 days and at a feed conversion efficiency higher than 50 per cent. Associated traits include a voracious appetite, initially to sustain the metabolic demands of rapid muscle growth but which persists after the animals reach physical maturity, and loss of synchrony between the development of the muscular, skeletal and cardiovascular systems (the animals outgrow their strength).

The most serious consequence of selection for birds that outgrow their strength is 'leg weakness', a euphemism for lameness and chronic pain. The aetiology of leg weakness in broilers is complex, ranging from failures of ossification (tibial dyschondroplasia) to septic arthritis associated with endemic bacteria. However, the main risk factor for all these separate conditions is selection for very rapid growth that puts the immature skeletal system under excessive stress, since these conditions are rare in birds selected for slower growth (slaughter at >70 days). The same applies to the cardiovascular problems (ascites and heart failure). Despite much high-level criticism (e.g. FAWC) of the breeding companies, recent independent evidence (Knowles *et al.* 2008) shows that the prevalence (though not the aetiology) of leg weakness has not significantly improved over the last 25 years.

The greatest welfare problems for broiler chickens arise as a direct consequence of breeding for rapid growth. They can be ameliorated through strict attention to husbandry, e.g. improved hatchery hygiene and feeding strategies designed to prevent overeating in early life. However the fast-growing strains are not fit for any purpose beyond growth to six weeks of age. It follows that the most effective way to improve the welfare of broiler chickens is to breed stronger, fitter birds. As ever, the most effective stimulus to progress has been public pressure operating through the major retailers. Many of the supermarkets in the UK market slow-growing strains of broilers (e.g. ISA S 457) promoted mainly on the valid basis of better welfare. In France, slow growing, free-range birds are marketed under the 'Label Rouge', mainly on the equally valid basis of better taste. One of the facts in favour of the promotion of better, kinder chicken is that higher welfare birds, by virtue of slower growth, really do taste better. However these birds are at present a very small minority. The vast majority of chickens killed for meat production on a world-wide basis (*c*.50,000 million birds per annum) are derived from fast-growing strains, nearly all originating from five international breeding companies. The policy of these breeding companies is strictly pragmatic: to develop a range of 'superior' products measured in terms of the specific requirements of their market. For most producers in most of the world the requirement is for fastest growth and highest feed conversion efficiency. As discussed in Chapter 3, it would be possible to prohibit certain strains on the basis of evidence that they 'were likely to cause suffering', to use the words of the UK Animal Welfare Act (2006). Realistically this is more likely to progress on a national (UK) or regional (EU) basis than through the agency of an international authority such as OIE. Legislators are more likely to be sympathetic to legislation for minimum production standards, e.g. a maximum of 20 birds/m^2 (500cm^2 per bird) at the time of slaughter, than for legislation that would bring them into direct conflict with the vested interests of the major breeding companies. Legislation to improve minimum production standards would improve the lot of the birds but should not be considered a satisfactory alternative to radical change in the phenotype.

Alternative, added value systems, such as free-range broilers, are generally satisfactory in terms of bird welfare and worthy of encouragement. However it should be pointed out that fast-growing birds (38–42 days to slaughter) make little use of free range. Slower growing strains (>70 days) do make active use of range, mostly in the pursuit of food. The message to the consumer is that the free-range label is not, on its own, a guarantee of high welfare; it must be accompanied by a guarantee that the birds are of a slower growing strain and reared to at least 70 days of age.

The most severe and prolonged welfare problems are experienced by the broiler breeders. If permitted to express their abnormal appetite for food they will eat themselves to an early death. Denied the expression of this most powerful of motivations they vent their frustration in aggression. It is extremely difficult to manage populations of broiler breeders and probably impossible to ensure their welfare. However these animals are extremely valuable to the breeding companies, which means that they get individual attention, unlike the 50,000 million birds reared per annum for poultry meat.

Eggs

The welfare of the laying hen has aroused more public concern and more action both in legislation and in the supermarket than any other aspect of intensive food production. The main commercial production systems are:

- Barren cages, with a minimum space allowance of 400–550 cm^2 per bird, depending on bird size and national legislation.
- Furnished cages, incorporating a nest box, perch and a substrate for comfort behaviour (dust bathing).
- Barns (enclosed colony units), accommodating flocks of birds in buildings containing nest boxes, perches and a substrate for comfort behaviour (dust bathing).
- Free-range units, involving accommodation similar to that in enclosed colony units *plus* space out of doors for foraging and other natural behaviours.

Welfare issues for laying hens have been rehearsed already in Chapter 3 (Table 3.2). The most obvious welfare problems for the laying hen (and the producers) are behavioural: aggression and feather pecking. The two are not the same. Aggression is relatively rare. Feather pecking is widespread and appears to be an antisocial distortion of natural investigative behaviour, since the beak is their main instrument for active investigation of the environment. However the consequences can be the same: injury and, in extreme cases, death. These problems tend to be greater in colony units than in barren cages containing only three to four birds, not least because the birds are presented with more opportunities and more space. In the barren cage, birds are denied almost any overt expression of behaviour but careful studies by ethologists have produced clear evidence of profound frustration (Nicol 2011).

The biggest physiological problem for the laying hen is osteoporosis, which greatly increases the risk of bone fractures. The fundamental cause is the continuous high demand for calcium for egg laying that cannot be met simply through the provision of calcium in the diet. This problem has been the subject of intensive study for many years but remains unresolved (Whitehead 2004). The degree of osteoporosis is exacerbated by inactivity, which means that birds confined in small barren cages have the weakest bones. However, birds in colony systems have more opportunities to injure themselves. Thus birds from cages are more likely to suffer broken bones during catching and transport at the end of lay. Birds in colony systems are more likely to experience fractures, especially of the keel bone (furculum) during the period of lay. The most immediate hazard would appear to be hard landings on perches, although this explanation is probably incomplete. The incidence of fractures increases greatly after the birds have been in lay for 20 weeks, which implies that osteoporosis, *per se*, is a necessary component of risk. It is also possible that some birds fracture their keel bones simply by flapping their wings. The degree of pain associated with fracture of the keel bone is unknown. We have good evidence that birds are highly sensitive to pain (Danbury *et al*. 1999). However, at present, there is no evidence of differences in behaviour between birds with and without healed fractures of the keel bone.

So where do we go from here? Clearly there has been a steady increase in demand for free-range eggs. In some parts of the UK this now exceeds 50 per cent. As I have written earlier, this has been driven more by public revulsion at the image of the spent hen from the battery cage than by careful analysis of all the factors affecting the welfare of laying hens. The question to be addressed at this stage is how well do the four alternatives listed above meet health and welfare needs as perceived by the birds.

Barren cages

These cages, stocked at 400–500 cm^2 per bird, i.e. as densely as possible, have been prohibited within the EU with effect from 2012, the justification being based almost entirely on the frustration of natural behaviour. It is fair to recall that cage systems evolved in the first place to reduce deaths from infectious disease (e.g. salmonellosis), parasitism (coccidiosis), predators and feather pecking. The decision to ban cages in the EU was based on the value judgement that the welfare cost of this extreme frustration of natural behaviour overrode all the other claimed advantages of the cage system measured in terms of health and hygiene.

Furnished cages

These have been developed on the basis of a great deal of scientific study of the behavioural needs of birds and their motivation to activities such as nesting, dust-bathing and comfort behaviours. Each cage may hold up to 60 birds. There is good evidence that the furnished cage providing a nest box, perch and dust bath or equivalent can meet most of these needs, which means that so far as the birds are concerned the furnished cage overcomes the problems of frustration while retaining the hygienic advantages of the battery cage (Appleby *et al*. 2010). The role of legislation for animal welfare is to set minimum standards. The development of premium, added value systems will be set by the free market. The EU specifications for the furnished cage will achieve significant improvements in bird welfare and deserve to be adopted on a much wider international basis, e.g. through the actions of OIE.

Barns

In these systems several thousand birds are held in climatically controlled houses furnished with nest boxes. Most of the floors are wire or slatted to permit passage of excreta but include an area of sand, straw or other suitable substrate for dust-bathing and foraging behaviour. Perches are usually though not invariably provided, often in several tiers, which increases the risk of keel damage. The former problems of infectious disease in colony units have largely been overcome through vaccination. The risks of deaths due to injury from other birds are greater than in cages. Furthermore in all large colonies there is always a risk that an outbreak of panic will lead to the mass death of birds killed in the crush. When assessed according to all Five Freedoms or twelve welfare criteria the barn system falls short of the furnished cage.

Free-range systems

These consist of a furnished barn plus access to the outdoors. In some units the entire floor of the barn is wire or slats but there is a verandah area between the barn and the pop holes to the outdoors where the birds may forage and dust-bathe. The disadvantages of the free-range system are similar to those of the barn system since all of the birds spend most of the time indoors and some of the birds spend all of their time indoors. The biggest point in favour of the system is that it permits birds greater freedom of choice. The quality of a free-range unit is greatly determined by the quality of the outdoor environment, as perceived by the birds. The modern commercial hen has been bred from jungle fowl, whose natural behaviour is to forage during the day under the cover of trees and shrubs to avoid predators in the air, and to rest at night on elevated perches to avoid predators on the ground. A well-designed unit provides overhead cover throughout the range and the birds make use of all the range. On a unit without overhead cover most of the birds stay close to the house, which increases the risk of parasitism and wastes most of the space nominally available to the animals.

Our team at Bristol made a comprehensive study of the welfare of laying hens on commercial free-range units operating to the standards of Freedom Foods (Whay *et al.* 2007). On nearly all units the health and welfare of the birds was very good. One surprising finding was that the incidence of aggression and injury in these flocks ranging in size from 3,000 to 13,000 birds was very low, considerably lower than that observed in barns, or indeed in experimental flocks of less than 100 birds. There was a real fear that the free-range principle could not be scaled up from backyard to major commercial operations without causing unacceptable behavioural problems. Clearly this is not a major concern. Indeed the reverse may be the case. Birds in very large flocks are unable to establish a pecking order. This may encourage them to be more discreet.

My personal opinion, based on considerable evidence, is that well designed and well managed free-range units can provide satisfactory standards of welfare for laying hens, at an extra cost to the consumer that need not exceed 20 per cent. In short, those consumers who currently buy free-range eggs in the belief that they come from happier birds can feel reasonably confident in their belief. Moreover those who currently do *not* buy free-range eggs may find it increasingly difficult to justify their decision on the grounds of cost.

Village chicken production

Chickens are a valuable source of high-quality protein in many rural areas in Africa and Asia. A comprehensive report on village chicken production system in rural Africa has been produced by the UN Food and Agriculture Organization (FAO 1998). Typical flock sizes vary from less than 10 up to around 100 birds, use indigenous, unimproved genotypes, and birds are not segregated by age or gender. Smaller flocks receive little or no supplemental feeding; they are expected to scavenge for themselves. Birds are kept for eggs and meat. Approximately 20 per cent of mature birds are likely to be in lay at any time. Birds eaten for meat are mature birds, including non-laying hens. Yields of eggs and meat from these birds are obviously very low in comparison with large intensive units. One sensible reason for this is they receive little or no food that could be used to feed the people. Other reasons include predation and infectious disease, especially Newcastle disease (ND) or fowl pest, which can cause mortality rates in excess of 50 per cent. Moreover, the larger the flock the greater the risk of infection and the higher the mortality. A programme of vaccination against ND, sponsored by international charities and implemented by trained ladies within local communities, has enabled families to increase their flock size from three to four birds up to numbers that can generate real income (20–100 birds).

Figure 6.1 illustrates the FAO step-wise approach to upgrading village chicken production from separate family units containing <10 birds to a semi-commercial level capable of generating better nutrition and real income. The first vital step involves better control of infectious disease accompanied by education in husbandry (feeding, hygiene, etc.). This first step requires external inputs. The next stage involves a careful programme of cross-breeding, sensitive to local resources and the local environment and resistant to pressure from the breeding companies. When chickens in family flocks are compelled to subsist for much of the year on what they can scavenge for themselves, the introduction of 'improved' strains can become self-defeating. In parallel with the development of healthy improved stock, there is a need to develop local producer and marketing co-operatives to explore the most efficient ways of converting local resources to food and income for those that need it most.

Food from other animals

There is not the space in this book to review major problems and possible solutions for all species of animals farmed for food. In addition to the species considered in this chapter, the UFAW publication *The Management and Welfare of Farm Animals* (2011) critically reviews current husbandry practices, and looks to the future for the following species: sheep, goats, red deer, game birds (e.g. pheasant and partridge), camelids (e.g. llama, alpaca), horses and donkeys (considered as working animals), ducks, ostrich and farmed fish. Table 6.5 presents a brief summary of the most important problems and possible husbandry solutions for

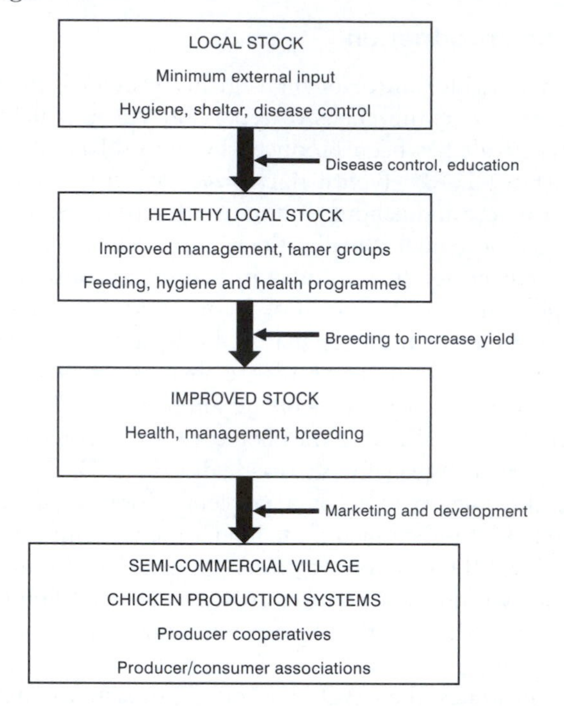

Figure 6.1 Step-wise improvement of village chicken production systems (from FAO 1998)

these animals, with the exception of working horses and donkeys, who are very important members of the community of total agriculture but not, in my book, considered as a source of good, kind food. One common feature of all the species listed in Table 6.5 is that none of them have been systematically subjected to the processes of hyper-intensification and confinement and the husbandry problems intrinsic to these systems. There are real problems for these species but, on the whole, they are not man-made.

Sheep

Traditionally the role of the shepherd was to move his flocks from pasture to pasture, protect them from predators, but otherwise leave them to fend for themselves. Sheep farming is an extensive exercise. In medieval times much of the wealth of Britain was based on the production of wool from sheep at grass. Today, except in special circumstances such as Merino sheep in Australia, the advent of synthetic fibres has eroded the value of wool to the point where the cost of shearing exceeds the value of the fleece. Today the sheep industry derives its income from the production of meat from lambs. Recalling once again Table 2.1, most of the feed and other resources are used to support the breeding population of adult ewes and rams. Lambs for slaughter drink mother's milk, eat grass and are killed between three and ten months of age.

Table 6.5 Major welfare problems and future solutions for 'minority' species farmed for food and fibre

	Current problems	*Future solutions*
Sheep	Internal and external parasites	Early diagnosis, effective prophylaxis
	Infectious foot rot	Eradication of pathogens, selection of resistant lines
	Neonatal mortality	Hygiene, management, quick dispersal
	Shearing	Genetic selection, 'easycare' sheep
Goats	Infectious foot rot	Eradication of pathogens, selection of resistant lines
	Unwanted male kids	Euthanasia, specialist outlets for kid meat
Red deer	Handling	Design of handling facilities
	Transport	Killing on site
	Harvesting antlers in velvet	Prohibition
Camelids	Handling	Design of handling facilities
	Immaturity at birth	Optimal environment, minimal intervention
Ducks	Water provision	Legislation
Ostrich	Handling, especially at slaughter	Design of handling facilities, training
Game birds	Housing of breeding birds	Legislation
Farmed fish	Welfare at slaughter	'Instantaneous loss of consciousness'

In the UK there is a stratified system of sheep production based on foundation stocks, of breeding ewes that fend for themselves in the hills and marginal lands, 'improved' flocks of more prolific hybrid ewes housed over winter, and terminal crosses between hybrid ewes and meaty sires to produce lambs reared wholly or mainly on fresh grass to produce high-quality meat. This system is ecologically sound since it makes best use of different land types but it does not generate much income. In particular, those who farm the foundation stocks on the hills could not survive without subsidy. There have been several attempts to create highly intensive systems of sheep production through increased prolificacy and year-round breeding in housed animals, but they have failed because the increased income from sale of meat does not justify the increased expenditure on capital and labour.

Some elements of the modest intensification of sheep production have brought about improvements in welfare, especially the housing and increased care of ewes at the time of lambing. Housing large numbers of ewes at this critical time inevitably increases the risk of infectious disease, but these problems can be managed by strict attention to hygiene at the time of parturition and the policy of getting ewes and lambs out to grass as soon as possible.

The most promising future for sheep is as constructive contributors to total agriculture, supported by the production of food, water management and carbon sequestration, the protection of amenities and the living environment through conservation grazing. These things will be discussed further in Chapter 8. This implies a trend to extensification rather than further intensification. This will lead to an increase in some risks to the health and welfare of the sheep, which can be overcome in part by reversion to the traditional practice of selection of sheep better able to look after themselves in specific environments. The reason why these days nearly all intensively reared pigs have the same phenotype is that nearly all are reared in identical, artificial environments; the reason why there are so many breeds of sheep is that the shepherds were unable to control the environment, so bred sheep to cope best with what they had and where they were.

There is however scope to improve the welfare of extensively reared sheep through the development of 'easy-care' flocks. This can involve, for example, introduction of genes from breeds such as the Wiltshire Horn to produce sheep that naturally shed their fleece in the spring. Other possibilities include the eradication of infectious foot rot through a combination of genetic selection and systematic eradication of carriers. The scientific approach to this will be discussed in the next chapter.

'Novel species': red deer, camelids, ostrich

I acknowledge that my use of the generic expression 'novel species' is a bit insular. Camels and South American camelids (e.g. llama, alpaca) have been domesticated for generations. Ostrich farming, principally for leather, with meat as a secondary product, is big business in Southern Africa. In Europe, farmed deer and ostrich are aimed at a high-value minority market. In order to preserve this image the producers themselves are extremely resistant to the concept of intensification and mass production. In the UK, camelids are mostly treated as pets. The main welfare problems for all these species arise from the fact that they have not been fully domesticated through generations of close contact with humans and selection for docility. Handling and movement of these animals can be extremely stressful to the animals and dangerous to the human handlers. These stresses can be reduced although not eliminated through design of structures that facilitate the movement and restraint of animals with minimal disturbance, risk of injury or direct contact with people (UFAW 2011). A more radical alternative is to eliminate the handling procedures altogether. Ruth Harrison, in a minority statement accompanying the FAWC Red Deer Report, recommended that farmed deer destined for human consumption should be killed by a shot from a free bullet while at a feeding trough on their home pasture. Since other deer do not demonstrate signs of distress when their shot neighbour falls to the ground, this would appear to be a humane option. Moreover there is no evidence to suggest that the risks to public health are greater when a deer is shot in the field than when it is slaughtered and eviscerated in a licensed abattoir. For wild deer shot by stalkers this is, of course, standard practice. This is one of very many examples where improvements to animal welfare and other features of good husbandry (such as the feeding of animals on surplus foods from supermarkets) are being restrained not by sound logic but by timid legislation. This will be a main theme of my final chapter.

7 Futures for animal science and technology

Everything is interesting; only some things are important.

(J. A. B. Smith)

The traditional definition of animal science has been that directed towards improved animal production through improved understanding of animal health, reproduction, growth rate, feed conversion efficiency and quality of product (meat, eggs, dairy products, etc.). This definition includes the branches of veterinary science directed to the health of farm animals and the safety of food from farm animals. Simply put, its primary purpose has been to advance the benefits we, the consumers, derive from farming animals to meet our needs at least cost to producers (defined as the agriculture industry). Today and tomorrow we are extending our definition of needs to incorporate values such as sustainability, pollution and animal welfare. We extend our definition of producers to include the farm animals and we make more effort to rate the benefits to us against the costs to them and to the living environment.

By this definition, animal science is both an applied science and one that is limited in scope. It is applied because it is directed to the 'better' use of animals to serve human kind. It is limited in scope because it relates only to a very small segment of animal biology, namely what happens within those animals used by man in agriculture. This contains much that is complex and fascinating. However it describes a rather small and rather inflexible link in the food chain. The major advances in the production of food for human (and animal) consumption have come from scientific and technological discoveries in crop science, relating primarily to productivity (including response to fertilisers) and disease resistance. The animals are merely links in the food chain between primary crops and the food that goes on our tables. Moreover it is inherently difficult to 'improve' animals, e.g. through genetic selection for accelerated growth or increased lactation, without compromising other traits related to evolutionary fitness to survive and reproduce.

There can be no doubt that the international animal production industry has succeeded in its aim of producing lots of food of consistently high quality at a reasonable price. It is however valid to ask 'How much of the progress to date is a direct consequence of animal science and how much would have happened anyway?'

Impacts of science and technology on food production from animals

Table 7.1 presents a matrix that examines the impact of science and technology on the management and welfare of the food animals. The elements of production are nutrition, reproduction, growth, lactation, health and welfare. The columns examine the extent to which they have, have not, or may be (in future) influenced by three drivers: practical technology, systems science and laboratory science. Each box is further subdivided according to whether the impact of the drivers has been *directive* (it has directly advanced production) or *remedial* (it has been used to address problems thrown up by technological progress). The examples presented in Table 7.1 are by no means comprehensive but selected for discussion of principles in regard to what animal science can do and cannot do, should do and should not do.

Animal nutrition

The science of animal nutrition rests on three central pillars:

- discovery of the requirements of animals for energy and specific nutrients; amino acids, minerals and vitamins for the purposes of maintenance, work, growth, pregnancy and lactation;
- understanding of the processes of digestion and metabolism that convert the chemicals in feeds to nutrients available for the functions listed above;
- understanding of the impact of diet on health, especially the health and normal function of the digestive tract.

The nutrient requirements of farm animals are now well documented, most comprehensively in the USA NRC (National Research Council) publications for all of the major farmed species. The gathering of all this information has taken more than a century of diligent scientific endeavour but today the job is done, give or take a little fine-tuning. Study of the physiology of digestion and metabolism has made a highly significant contribution to the practice of animal nutrition by revealing the factors that determine the efficiency with which the chemical component of feeds are converted into nutrients available for metabolism (feed conversion efficiency). This work too has been enormously valuable and the results are good enough for most practical purposes. This should be cause for congratulations to previous generations of nutrition scientists. However I would not recommend it to a young graduate seeking a future in animal science. The third pillar, impact of diet on health, especially the health of the digestive tract, is the least understood, and therefore the most important topic for the future. It is also the most interesting although, in an applied science, interesting is not enough. Throughout my career I have sought to remember the advice of J. A. B. Smith, my first director and supervisor of my PhD thesis: 'everything is interesting; only some things are important'.

Table 7.1 Examples of the impact of technology, systems science and laboratory science on elements of animal production, (adapted from Webster 2002)

	Technology	*Systems science*	*Laboratory science*
Nutrition			
Directive	Crop improvement ++	Feed evaluation +++	GE, crops ++
	Feed processing +++	Nutrient requirements +	
Remedial	n.s.	Feed evaluation +	Deficiency disorders ++
Reproduction			
Directive	AI +++	n.s.	Endocrinology +
	MOET +		Molecular biology n.s.
	Pregnancy diagnosis ++		Cloning n.s.
Remedial		Ethology +	Endocrinology ++
Growth			
Directive	Genetic selection +++	Population genetics +	Marker-assisted selection +
			GE, animals – –
			Endocrinology +*
Remedial	Genetic selection +		Bone & matrix biology +

	Technology	*Systems science*	*Laboratory science*
Lactation			
Directive	Genetic selection +++	Population genetics +	Marker-assisted selection +
	AI +++	Nutrition +++	Endocrinology +*
	MOET +		GE & cloning n.s.
Remedial	Genetic selection +		
	Milking technology+		
Health			
Directive	Intensification – – –/(+)	n.s.	Vaccines, chemotherapeutics +++
Remedial	Hygiene +++	Epidemiology ++	Vaccines, chemotherapeutics +++
	Genetic selection +	HACCP ++	Genomics, disease resistance +++
Welfare			
Directive	Genetic selection +	n.s.	n.s.
Remedial	Humane slaughter ++	Ethology +	Neurobiology n.s.
		Motivation analysis ++	Bone & matrix biology +

Key: AI = artificial insemination, GE = genetic engineering, HACCP = hazard analysis and critical control points, MOET = multiple ovulation and embryo transfer, n.s. = no significant effect.

Plant science and technology has, I repeat, made great progress towards increasing crop yields and reducing losses from pests and diseases. However I shall not discuss these things further as they are outside the scope of this chapter. Plant science has, to date, had a limited impact on the nutritive value of individual dietary constituents (e.g. cereals and grasses, oilseeds). There have been some advances through genetic selection, e.g. the reduction of anti-nutrient factors in oilseeds (soya and rape) and the breeding of high-sugar grasses for ruminants, either grazed or conserved as silage. In both these examples this makes practical sense. Oilseeds are an almost obligatory protein source for simple-stomached animals. Grass is an almost obligatory and sometimes the only feed source for ruminants. However the most highly nutritious grass is an unbalanced feed, containing relatively too much rumen degradable protein relative to fermentable energy for optimal rumen fermentation. The creation of better-balanced 'high-sugar' grasses by genetic selection within existing cultivars or through genetic engineering ranks high on the list of things to do (Gardner *et al.* 1997). For the most part however there is little point in using high science to manipulate the composition of a single dietary source in the effort to make it into the perfect feed. It makes more sense to make an appropriate selection from available feed sources to formulate a balanced, least-cost diet.

One of the greatest contributions to improved nutrition, especially for ruminants, has been through the application of systems science. Feed evaluation, especially for ruminants, has built on foundations of fundamental knowledge in a number of disciplines: rumen microbiology, fermentation dynamics, physiology of digestion and N recycling (etc.). Simply expressed, modern diets for high-producing ruminants are designed first to provide a balance of nutrients for the rumen microbes, second to calculate the manufacture of nutrients from microbial metabolism (volatile fatty acids and microbial protein), and finally to calculate the requirement for additional nutrients that bypass the rumen (Fox *et al.* 1992; Alderman and Cottrill 1993). It would have been impossible for cattle breeders to have increased lactation yields from 5,000 to 10,000 litres per lactation without the parallel development of 'total mixed rations' designed to achieve very high nutrient intakes within the constraints of appetite and without compromise to stable fermentation in the rumen. In Table 7.1, feed evaluation is also identified as *remedial.* There is no doubt that improved ration formulation has significantly reduced the prevalence of digestive disorders in both beef and dairy cattle fed high-energy rations. I have elsewhere (pp. 126–8) drawn attention to serious welfare problems in high-yielding dairy cows. However, the risks attributable to feed quality *per se* have declined. I consider that improved ration formulation, based on improved feed evaluation, constitutes one of the most important contributions of animal science to better food production from ruminant animals, measured both in terms of productivity and welfare.

Nutrition, in common with most animal science, is a quantitative science: i.e. it rarely adds to knowledge through new discovery but through progressive but progressively diminishing improvements in efficiency achieved by changing inputs (in this case feeds) or by changing the processing of feeds into available nutrients, either by digestion and metabolism within the animal, or by processing before presentation to the animal. In any quantitative science it is necessary for the sponsors acting for society at large to decide when enough is enough: when the benefit, measured in terms of increased efficiency of production, ceases to justify the cost, measured both in terms of the financial costs of the research and the potential negative effects on the animals and the environment. I suggest that we already know enough for practical purposes about the efficiency of digestion and metabolism of energy and protein in conventional feeds for both ruminant and simple-stomached animals.

There is scope for future science and technology designed to improve the quality of feeds at the point of presentation through increases in digestibility and reduction in digestive disorders. Some of these are presented in Table 7.2. For simple-stomached animals such as pigs and poultry (and calves reared for white veal) it is sometimes economically worthwhile to improve the digestibility of starches through cooking or micronising (as in breakfast cereals for humans). This is important with relatively indigestible maize starch. It is critical for piglets that have been weaned too early before their starch-digesting enzymes have sufficiently developed. Increasing starch digestibility is generally contraindicated for adult ruminants because it predisposes to ruminal acidosis. Diets for early-weaned calves with immature rumens may contain flaked (micronized) maize.

Table 7.2 Potential benefits to digestibility and nutritive value of feeds from feed processing and feed additives

	Pigs and poultry	*Ruminants*
Feed processing		
Increase starch digestibility	Cooking, micronising ++	Contraindicated
Increase fibre digestibility	n.s.	NaOH treatment +
Increase protein availability	Reduce antinutrients in oilseeds ++	Reduce degradability in rumen +/–
Feed additives		
Antibiotics	Effective, + but widely banned	Generally contraindicated
Probiotics	Unproven (n.s.?)	n.s.
Prebiotics		
Rumen 'stabilisers'	Not applicable	Buffers (sodium bicarbonate) ++
		Ionophores + but widely banned
		Lactate using bacteria

n.s. = no significant benefit

Plant fibre describes the constituents of the plant cell wall, a structural matrix of cellulose, hemicellulose and lignin. Cellulose and hemicellulose can only be digested through fermentation by microorganisms (bacteria and fungi) present in some parts of the gut of all animals. Lignin is indigestible. Ruminants and equines can digest much more fibre than simple-stomached animals, not through any fundamental differences in microbes and microbial metabolism but simply because the space and number of microorganisms available for fermentation in the rumen and large intestine is so much greater. The digestibility of highly fibrous diets for ruminants (e.g. barley straw) can be improved by treatment with caustics (e.g. NaOH). This does not significantly alter the chemical composition but opens the structural matrix of the plant wall, thus permitting the microorganisms greater access to the cellulose and hemicellulose. The scope for manipulating the quality of fibre to increase digestibility in simple-stomached animals is severely constrained by the anatomy of the gut. The rate of microbial fermentation can be increased, but the time and space for the process will always be very small.

New science has entered this arena with the creation of the 'enviropig'. This animal has been genetically engineered to secrete the enzyme phytase in the saliva (Streiffer and Ortiz 2010). There are several good reasons for incorporating relatively high amounts of fibre in diets for pigs, especially pregnant sows. They are complementary diets, relatively cheap, provide oral satisfaction and, up to a point, they are compatible with satisfactory performance. However in simple-stomached animals fibrous feeds such as wheat bran bind phosphorous (P, both dietary and circulating) into complex, poorly digestible molecules within the gut, with the result that large amounts of P are excreted in the faeces. This increases dietary P requirement, which further increases P excretion: a major source of pollution from intensive pig farms. This is not a problem for ruminants that have phytase-secreting bacteria in the rumen.

When we identify a threat to the environment where the cause is indisputably intensive agriculture, but the mechanism is understood and prospects for a scientific solution are good, we see this as an opportunity. Environmentalists call for something to be done, scientists call for something to be done *by me*, government ministers call for action to allow them to say that something *is* being done. Examining the issue from first principles, one can offer three options for action.

- Feed less fibre to pigs.
- Pre-treat high-fibre diets for pigs with phytases to reduce the P-binding effect.
- Genetically engineer phytase into the saliva of the pig itself.

The first solution is easy and too obvious to attract the attention of scientists and their sponsors. The second solution is a straightforward matter of technology. It can be done. The extent to which it will be done will be determined by simple economics: increased feed cost set against increased cost or reduced opportunity incurred as a result of staying within quotas for P output. The third option is clearly the most attractive to animal scientists who have the resources to genetically modify farm animals and are searching for some real use to which it might be put. It is the most attractive to politicians anxious to be seen to support green science. Hence, the 'enviropig', a pig genetically engineered to secrete phytase in its saliva. As a scientific experiment this has been a success since P excretion can be reduced by 50 per cent in GE pigs. This research was funded by government but has not been taken up by the private sector. One good reason is undoubtedly an awareness of public antipathy to GE animals and food from GE animals. A sounder reason is that most folk in the pig industry view it as an unnecessarily complex and expensive solution to a relatively simple problem: not so much a sledgehammer to crack a nut as a sat-nav to find your way to the kitchen. I offer this as a classic example of the highly interesting but supremely unimportant.

Probably the most active area of scientific research within the area of animal nutrition, now and for the foreseeable future, concerns the development of feed additives designed to maximise performance in ruminants fed high-starch:low-fibre diets, typically beef cattle fattened on feedlots in North America. The aim of these additives is to stabilise rumen fermentation and prevent recurrent rumen acidosis, with excessive production of lactic acid, that carries a severe cost to the farmer in terms of reduced appetite and growth rate, and a severe cost to the animals through the primary discomfort of rumen indigestion and a number of pathological sequelae including parakeratosis of the rumen epithelium, laminitis and liver abscesses. Given the enormous size of the cattle feedlot industry there are big potential gains to be made from the development and patenting of a feed additive that can, at relatively low cost, significantly increase both growth rate and feed conversion efficiency: hence the enormous and sustained investment of time and money into research in this area. Table 7.2 lists some of the more effective approaches. Feeding buffers (heroic quantities of sodium bicarbonate) is very effective but unappealing to entrepreneurs since it cannot be patented. Ionophores (e.g. monensin) are effective in reducing rumen acidosis, and also reduce methane production. There are some problems associated with the feeding of monensin to ruminants but, on balance, and used with care, I believe that monensin can do more good than harm. However its use is banned in Europe through application of a precautionary principle that has rather little foundation in science. The other option (one of many) listed in Table 7.2 is the incorporation of bacteria able to utilise and so reduce the concentration of lactic acid in the rumen. This approach is theoretically attractive but it remains to be seen how well these bacteria are able to compete within the microbiological free-for-all of the rumen.

It is essential to stress that the action of all these 'rumen stabilisers' is remedial. If they work, the way in which they work is by reducing the severity and risk of profound rumen indigestion and consequent pathology such as liver abscesses, when cattle are fed unhealthy diets. Developers and promoters claim that these additives 'improve' performance in feedlot cattle. What they actually mean is that when cattle are fattened unnaturally fast on unnatural rations, these additives may help to ensure that fewer fall sick.

The last two categories of potential feed additive listed in Table 7.1 are antibiotics and probiotics. It is undeniable that the routine incorporation of antibiotics into the diet does increase growth rate and feed conversion efficiency in simple-stomached animals such as pigs. On commercial units much of this can be attributed to a reduction in chronic low-grade infection, including respiratory disease and post-weaning enteritis, and on this basis it is valid to condemn the practice for helping to sustain profitability in a fundamentally unhealthy system. In fact, the story is not so simple. Broad-spectrum antibiotics can improve feed conversion efficiency in growing pigs reared in hyper hygienic, biosecure, minimal disease environments and showing no evidence of clinical disease. They do this by destroying a high proportion of the bacteria in the gut, not only potential pathogens but also the commensal bacteria naturally found in both the small and large intestines. Some of these bacteria compete for nutrients; others in the large intestine contribute to nutrient supply through fermentation of fibre. Those in the small intestine contribute to a stable microenvironment that inhibits the development of pathogens; i.e. they have a *probiotic* effect. However the stability of the microenvironment also depends on a great deal of immunological activity within the cells that line the small intestine (especially the terminal ileum). This is an entirely healthy adaptation in an animal exposed to a normal environment. However it is energetically expensive. The tissues of the small intestine are amongst the most metabolically active cells in the body. The routine administration of oral antibiotics to young pigs reared in a near sterile environment can increase the efficiency of utilization of ME (see pp. 27–8) by reducing the metabolic rate of the intestinal tissues. This is however a highly dangerous strategy since it effectively switches off the natural immunological defence mechanisms so that if a pathogen does get access the results can be catastrophic.

The main reason for prohibiting the routine use of antibiotics in diets for farm animals has been the risk of developing antibiotic resistance in bacteria that can cause serious disease in humans. The emergence of resistant strains of dangerous bacteria such as methacillin-resistant *Staph. aureus* (MRSA) and *Clostridium difficile* creates present and future problems in human medicine, although it is probable that the antibiotics previously permitted for routine incorporation into animal feeds have made little contribution to these increased risks. As discussed in Chapter 3, the ban may contribute to *increased* risk by encouraging unscrupulous producers and their equally unscrupulous veterinarians to prescribe for routine, nominally therapeutic use antibiotics that are essential for human medicine.

The routine incorporation of broad-spectrum antibiotics into feeds for ruminants is contraindicated for the obvious reason that they would inhibit the essential microflora of the rumen. In the past, 'growth-promoter' antibiotics were commonly incorporated into milk replacer feeds for artificially reared calves prior to weaning and calves reared for white veal on all, or nearly-all liquid diets. The prevalence of enteric disease in white veal calves is high (frequently >30 per cent) and it is common to control disease in these animals by metaphylaxis; i.e. when a few get sick, treat the lot. White veal production, despite new regulations within regard to diet and housing, remains one of the greatest abuses of good husbandry and serious risks to health and welfare of calves and humans alike.

In the 1980s certain permitted antibiotics were incorporated into commercial milk-replacer diets for artificially reared calves in the UK as a sales gimmick, in much the same cynical way as scientific-sounding additives like chlorophyll and provitamins have been used to promote new brands of toothpaste and shampoo. These antibiotics had limited action only against a small range of Gram-positive organisms. They had no effect on Gram-negative organisms such as *E. coli* and the *Salmonella* species responsible for most enteric disease in calves (and humans). Moreover, by inhibiting Gram-positive organisms and some of the commensal bacteria they probably *increased* the risk of enteritis in the calves (this is unproven). When I asked the manufacturers why they persisted in including a feed additive that was useless at best, and probably risky, their reply was 'If it doesn't say antibiotic on the bag, the farmers won't buy it'. The moral of this story is that there are times when we need legislation to protect us from ourselves.

In recent years there has been increased interest in the use of probiotics and prebiotics, driven largely by the natural desire of scientists and entrepreneurs to seek personal gain. The argument for probiotics is based on the scientifically sound premise (discussed above) that the health of the digestive tract is preserved by a healthy balance of commensal bacteria. Probiotics therefore consist of live cultures of organisms (e.g. *lactobacillus acidophilus*) whose aim is to escape acid digestion in the stomach and populate the terminal ileum with responsible colonists, thereby keeping out the criminal element. This is fine in theory. However I remain sceptical as to the likely impact of probiotics on animal and human health, while conceding that in the interim several organisations might make considerable sums of money from their sale. My reason for this is that the microbial ecology of the terminal ileum is constantly in a state of dynamic flux: population numbers of specific microorganisms can explode and collapse within days. There is, as yet, little convincing evidence that probiotic cultures are any less unstable.

An alternative approach is through the use of *prebiotics*. These are compounds designed to encourage the natural development of healthy bacteria in the gut. This is, of course, one of the functions of a normal healthy diet. At best therefore pro- and prebiotics may be viewed as potentially remedial feed additives designed to reduce the risks associated with feeding too much of the wrong sort of feed at the wrong time. At worst they may be viewed as a device for persuading people to spend money to resolve a problem that they should not have got into in the first place.

Breeding and reproduction

In the matters of farm animal breeding and reproduction, the aims of applied science and technology are:

- to breed as many animals as possible: maximise annual reproductive rate in breeding females;
- to breed 'better' animals: increase genetic potential for meat, milk and egg production.

In the case of meat production, the greatest gains in overall efficiency can be achieved by increasing the number of offspring produced by breeding females in order to minimise the overhead costs of maintaining the breeding generation relative to that of the slaughter generation. Cows are physiologically constrained to produce one calf per year. In the case of beef production from suckler cows approximately 50 per cent of feed is consumed by the breeding generation. In the case of broiler breeders producing 240 live chicks/year, the proportion of feed eaten by the breeding birds is only 5 per cent (Table 2.1). In the case of milk production from cows, reproductive rate remains fixed at one calf per year but productivity has doubled from 5,000 to 10,000 litres/year through genetic selection for increased yields supported by reproductive technology (e.g. the long-term storage of frozen semen and, to a lesser extent, frozen embryos) designed to facilitate rapid world-wide distribution of the most highly desired genetic material.

Increasing reproductive rate

The greatest increase in production through increased reproductive rate has been achieved in commercial poultry, whether for the sale of sterile eggs for human consumption, or day-old chicks for meat production. This has not required any high science; merely an understanding of the normal physiological processes of egg laying in birds supported by conventional procedures of genetic selection. In the wild, the jungle hen will, at a favourable time of year, lay one egg per day until she has completed a clutch of six to nine eggs that will have been fertilised in the reproductive tract by a passing cockerel. She will then cease laying, brood the eggs and rear her chicks to the best of her considerable ability. The basic triggers for egg laying have been known to us for centuries. Ovulation is independent of mating, onset of ovulation is triggered by light, ovulation ceases when the hen recognises (unconsciously) that the clutch is big enough. Place hens in a confinement building with a controlled lighting pattern (typically 15 hours light, 9 hours dark) that eliminates seasonal change, routinely remove the eggs and they will continue to lay one egg per day (almost) for months on end. In fact the average interval between eggs is about 25 hours, and eggs are usually laid within the first four hours of the period of light so the natural tendency is to skip the occasional day to re-establish the rhythm. Research (Fitzsimmon and

Newcombe 1991) has shown that this can be overcome by creating an artificially extended 'day' of 14 hours light, 14 hours dark (14L:14D). This is, of course, only possible in a total confinement system that allows the birds no access to natural daylight and is currently banned in the UK on the grounds that it interferes with the natural diurnal rhythm of the birds. There is little direct evidence for this. However the commercial advantages of a 14L:14D cycle seldom justify the outlay. Egg size and shell quality may increase but egg numbers are likely to be less. In a free-range system it is obviously impossible to operate anything other than a normal 24-hour cycle. However egg laying can be sustained throughout most of the year by maintaining a 15L:9D cycle within the houses.

Pigs

Mammals only become pregnant after a fertile mating and the duration of pregnancy is fixed. Thus the only opportunities for increasing reproductive rate are to increase litter size and/or reduce the interval between subsequent conceptions. The most effective way to increase the reproductive performance of breeding mammals is likely to be through reduction in neonatal mortality.

Commercial pig breeding takes the form of a 'pyramid' (Figure 7.1). At the top of the pyramid are nucleus herds of purebred sows selected to breed true for specified traits such as growth rate, carcass composition and prolificacy. The multiplication herds are those in which different purebred lines are cross-bred to produce the mothers for the production herds. Cross-bred animals carry hybrid vigour, which is particularly important in relation to mothering ability and breeding success. Hybrid sows for intensive systems may typically be crosses between purebred Landrace (contributing especially carcass quality) and Large White (or Yorkshire, contributing size and prolificacy). Other breeds, such as the British Saddleback (or Hampshire), may be used to produce hybrid sows suitable for outdoor units. Most significant increases in prolificacy (measured in numbers of piglets delivered alive or dead) may be achieved by cross-breeding with the highly prolific but very fat Meishan breed. This has not proved popular since the increase that can be achieved in numbers of pigs raised to weaning is unlikely to offset the reductions in growth rate, carcass quality (leanness) and feed conversion efficiency.

Scientists have located the position of the genes linked to productivity in the Meishan pig, which offers the prospect of accelerated selection for productivity either though marker-assisted selection or genetic engineering. Marker assisted selection (MAS: Dekkers 2004) is an indirect selection process for a trait of interest (in this case prolificacy), not based on the trait itself, but on a genetic marker (allele) linked to it. The assumption is that the linked allele associates with the determining gene or genes. In many cases MAS is used to identify a quantitative trait locus (QTL) of interest. QTLs are stretches of DNA containing or linked to the genes that underlie a quantitative trait. Increasing the rate of genetic progress through application of MAS involves good science in the identification of potential markers supported by normal breeding

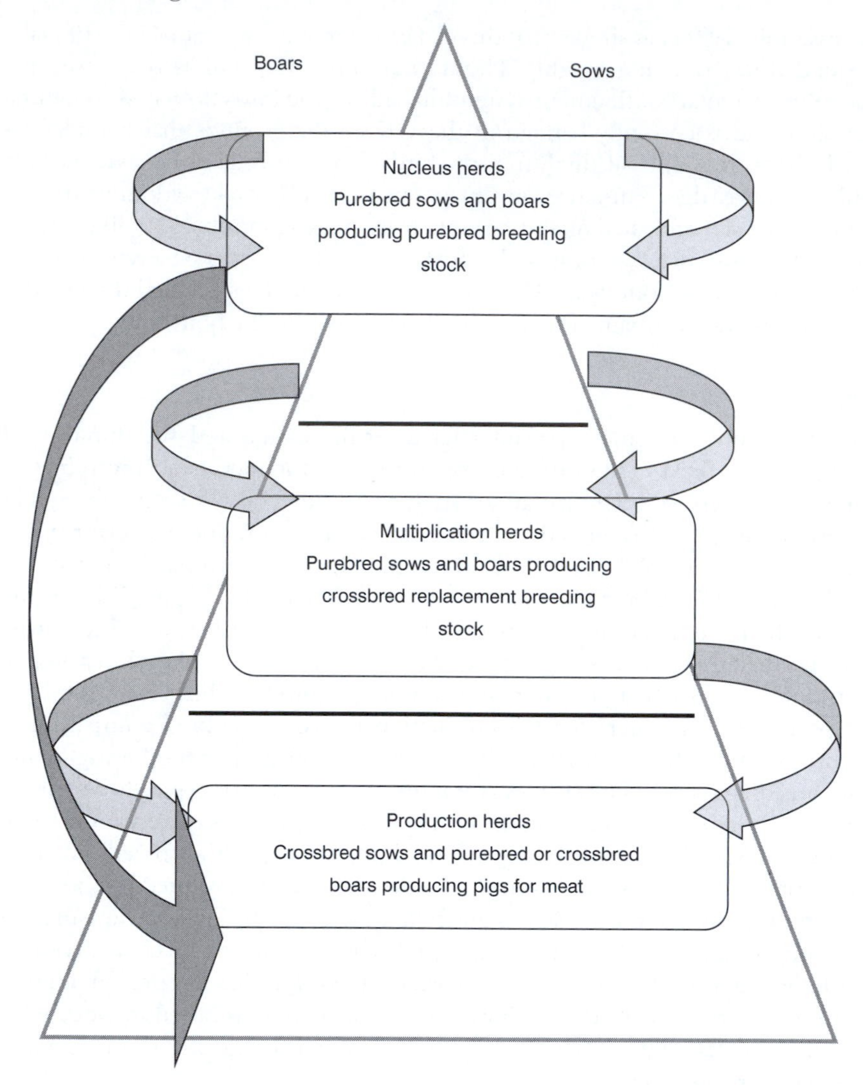

Figure 7.1 The typical breeding 'pyramid' in commercial pig production

practices. Whether the improvement in the selected trait turns out to be real, commercially significant and unaccompanied by undesirable associated traits can only be discovered by doing the experiment. In principle, this has to be an entirely acceptable and potentially valuable area of applied animal science.

Having identified genes linked to prolificacy it is theoretically possible to create genetically modified pigs through the technique of gene insertion with the view to increasing prolificacy. As always this approach would be likely to meet with public disapproval, measured in terms of reduced demand for the product. Moreover, as with the enviropig, it really doesn't make much practical sense. Increasing numbers of piglets born almost inevitably increases the number that fail to survive to weaning. At present, the greater priority for the pig industry is to reduce piglet mortality.

Pigs in confinement units can breed throughout the year. Sows normally come into oestrus five to seven days after weaning their previous litter of piglets. One can, in theory, maximise productivity, measured as piglets per sow per year, by weaning as early as possible. Measured simply in terms of guaranteed return to oestrus, the time from parturition to weaning can be reduced to about 17 days and many producers have attempted to push their sows to this limit. As so often however, the push to maximise output according to one simple crude measure generated a stack of new problems and new costs, most conspicuously post-weaning enteritis and the need for high-cost, low-welfare accommodation (flat-deck cages) for immature piglets. These things were discussed in Chapter 3. In the context of this chapter the message is, once again, that in the matter of reproduction it is especially hard to improve on nature.

Sheep

On a world-wide basis, most sheep farming is a very extensive business: the breeding flock is left to forage for itself, but given protection from predators as necessary by shepherds, often with guard dogs. The large white, woolly-looking Pyrenean mountain dog could be mistaken for a sheep at a distance but would present a nasty surprise to a wolf at close quarters. In this extensive state, and in temperate latitudes, ewes are short-day breeders. They start to display oestrus activity in the autumn, so that their lambs are born after a five-month gestation in spring, to exploit the new growth of grass. This is a highly efficient, sustainable system supported almost entirely by complementary feeds. However, like most low input–low output systems it doesn't generate much income simply from the sale of meat. Moreover, in the eyes of the reproductive physiologist, the production of only one lamb per ewe per year appears too leisurely by half.

Theoretically the reproductive performance of the ewe can be increased by increasing conception rate and by reducing the interval between lambings. Increasing conception rate in ewes can be achieved by exploitation of existing genetic variation supported by appropriate nutrition. Breeds such as the Blue-faced Leicester and Finnish Landrace are particularly prolific, giving birth to a high proportion of twins and triplets. Conception rate in all breeds can be increased by improving ewe nutrition on 'flushing' pastures prior to the introduction of the breeding rams. The stratified sheep industry in UK operates on a rather similar basis to the pig industry as illustrated in Figure 7.1. Nucleus flocks of purebred hardy ewes in the hills produce lambs at a rate not much in excess of 100 per cent. Some of these ewes are sent to multiplier units and mated with 'improver' rams (e.g. Blue-faced Leicester) to produce flocks of more prolific cross-bred ewes that can achieve lambing percentages close to 200 per cent. These cross-bred ewes are then sold on to production units on lowland farms where they are mated with meaty breeds of ram (e.g. Suffolk, Texel) to produce as many high-quality lambs as possible from grass, with minimal dependence on purchased feeds. Inevitably the prolific cross-bred ewes require more attention to nutrition, housing and management at lambing; it has become a high-input/high-output system. The production of twins carries a high risk of metabolic disease (pregnancy ketosis) unless the ewes receive an adequate supply of starchy concentrates in late pregnancy. The metabolic demands of producing triplets is proportionately greater and presents the further complication that sheep have only two teats, which means that at least one triplet is unlikely to thrive unless fostered onto a mother with a teat to spare.

All this has been achieved without recourse to science or high technology. However the sheep industry has profited enormously from one development in reproduction technology, namely the use of ultrasound scanners to diagnose pregnancy and count the number of embryos in the uterus. This application of technology is highly cost-effective since it enables the farmer to group ewes in late pregnancy according to the number of lambs they are carrying (0–3) and feed precise amounts of purchased concentrates to meet their specific requirements.

Gene(s) associated with prolificacy in sheep have been identified, the most well-known being the Booroola gene, which has been associated with crops of four to six lambs. As with pigs, the potential exists to increase prolificacy further in ewes either through conventional selection, marker-assisted selection or gene insertion. At present, however, it would appear the potential benefits, measured in terms of increased lamb sales, are unlikely to justify the costs to shepherds and sheep.

The other theoretical way to increase prolificacy in sheep is by decreasing the lambing interval. Since the duration of pregnancy is five months and sheep can exhibit oestrus while lactating, it is possible to achieve two lamb crops per year from breeds with an extended breeding season. About 30 years ago, when increased productivity at almost any cost was the big motivating factor for animal scientists, there was a move to develop highly intensive sheep production systems on similar lines to those operating for pigs. Cross-bred ewes, derived from Dorset Horns (extended breeding season) and Finnish Landrace (high prolificacy) were permanently housed under controlled lighting and presented with a high-energy, high-starch diet. The exercise proved to be a success according to the parameters of the experiment since the ewes produced four to five lambs per year. By any other measure it was a flop. The costs of housing and feeding were excessively high relative to income from sale of rather poor-quality lambs. Moreover the welfare and productive life of the ewes were highly unsatisfactory. Many had to be culled after only two to three lambings, typically as a consequence of mechanical collapse in their udders, some of which dragged on the ground.

There can be no doubt that the intelligent application of good husbandry to the modest intensification of the sheep industry has met with some real success. Systems based on ewe flocks that produce two lambs per year are fed mainly on pasture and conserved forage supplemented by controlled, limited amounts of concentrates, are kept in good housing and, with careful attention at lambing, do better than extensive sheep on hills and marginal pastures, measured both in terms of productivity and welfare. However the only significant contribution from science and technology has been the advent of the ultrasound scanning. Physiologists have produced other potential aids to improved reproductive performance in sheep, including artificial insemination and hormone-induced synchronisation of oestrus. These techniques have considerable value in cattle breeding (see below) but have not caught on in the sheep industry. Natural breeding usually makes the most sense. Rams are easy to manage (compared with bulls). AI is expensive and less successful than with cattle. The pressure to select for highly heritable traits in the slaughter generation, such as growth rate and feed conversion efficiency, is much less than in pigs and poultry since most of the costs relate to the maintenance of the breeding generation (Table 2.1) where the benefits of selection are less since heritability for reproductive traits is low relative to the impact of hybrid vigour. Synchronisation of oestrus only makes sense when combined with artificial insemination. When ewes are running free with the rams, it makes absolutely no sense for them all to come on heat at once.

Cattle

The duration of pregnancy in cattle is nine months. Cows are unlikely to return to oestrus and conceive until 30–40 days post-calving. Thus, in a seasonal, pastoral system, it usually makes sense for the cows to produce one calf per year in the optimal season. For beef and dairy cows on pastoral systems, this is likely to coincide with peak growth of spring grass, to meet peak demand for nutrients in early lactation. In more intensive systems, where cows are housed and fed conserved forage and concentrates, optimal calving time is more likely to be determined by seasonal variation in the price of milk or beef than by variation in feed supply. In cattle there is no practical scope for increasing prolificacy. When a cow conceives twin calves of opposite sex there is a high risk that the calves will be intersex 'Freemartins' due to mixing of the foetal circulations *in utero*. There is also good evidence that twinning increases the risk of post-partum disorders in the cows including retained placenta and increased risk of subsequent infertility. Inevitably some animal scientists have carried out experiments to increase twinning rates in beef cattle on the assumption that two calves per cow should yield more profit than one. Happily this idea has not been taken up by the industry. The obvious costs to the cows were just too high.

So the best we can expect from cows is one calf per year. For the traditional dairy cow at pasture for much of the year, this was entirely satisfactory. As milk yield declined with advancing pregnancy the cows were dried off two to three months before the next calf was due and given time to recover condition. For the modern higher yielding dairy cow on the intensive grass pastures of New Zealand, optimal calving time still coincides with the arrival of the best spring grass. For the modern high genetic merit Holstein cow, spending most of her life in a barn, there is no real need to match calving to the season. Indeed a cow with a total lactation yield of 10,000 litres and a calving interval of one year may well still be giving over 30 litres per day after 10 months when she needs to be dried off prior to her next calving. This seems wasteful, is almost certainly uncomfortable and carries an increased risk of mastitis.

For many herds of dairy and beef cows, managed to maximise output of milk or beef relative to input of feed, the main production disorder is infertility through failure to conceive while in early lactation. The aetiology of infertility in cows is complex and has been exhaustively researched. In the absence of a bull, one of the greatest risk factors is failure to recognise oestrus and present the cow for AI at the right time. The greatest of the physiological risk factors is loss of body condition in early lactation through failure of ME supply to meet the energy demands of lactation. This is associated with abnormalities of endocrine function that are, in theory, amenable to hormone therapy. However the most effective approach is through proper nutrition both in early lactation and in the transition period prior to calving. Increasing calving interval from say 365 to 400 days due to failure to conceive in the first three months of lactation is not necessarily a problem for high-yielding Holsteins but it can seriously compromise productivity in dairy and beef herds on pastoral systems. For the high-yielding Holstein the big problem does not arise from a modest extension of the calving interval but from cows that fail to conceive at all and have to be culled after three lactations or less.

There has been considerable scientific interest in the case for extending lactation length in high-yielding, barn-fed dairy cows (Arbel *et al.* 2001). Theoretically, optimum lactation length would appear to be about 18 months on the basis that a cycle of 16 months lactating at >30 litres/day + 2 months dry is more efficient than 10 months lactating plus 2 months dry. There is a further potential advantage in that most of the health and welfare problems of dairy cows occur in the first three months of lactation. Increasing the ratio of safe to risky months from 7:3 to 13:3 looks like a good idea. In the light of the previous paragraph one might also imagine that planned insemination of cows nine months after calving would be associated with higher fertility than at two months when they are in negative energy balance. Experience to date however suggests that this is not the case. This is disappointing but it may be solved by future research. I believe that for many high-output dairy cows, extended lactations could prove to be a better idea.

The most striking and economically successful results of breeding science and technology have been those directed towards the rapid dissemination of superior genotypes through the international dairy industry. Just what does and does not constitute 'superior' has been discussed elsewhere (pp. 128–9). This section considers the technology that has made it possible. The single most effective technology has been the successful development of artificial insemination based on frozen semen that can be stored indefinitely. This has enabled semen from bulls proven superior through progeny testing to be used on many thousands of cows throughout the world.

In recent years, it has become possible to produce sexed semen with a probability of fertilisation close to that of conventional semen (about 65 per cent to first service in fertile cows). The technique involves separation of X- and Y- chromosomes based on the staining of DNA and flow cytometry (Johnson 1995). This can make a significant contribution to better husbandry by ensuring that the number of female calves sired by semen from a superior dairy bull matches the requirement for replacement heifers, thus reducing the number of unwanted low-value males likely to be slaughtered shortly after birth or, which may be worse, shipped into white veal production units. At present many farmers are restricting the use of sexed semen to heifers, where fertility is expected to be high. There is some evidence that fertility to sexed semen may be unsatisfactory in older dairy cows. However this may be more a consequence of reduced fertility in the cows than in the semen.

The practice of AI requires human intervention both for insemination and diagnosis. Generally speaking we are less effective at these operations than the bulls themselves. Moreover we are able to devote less time to the work. Hormonally induced synchronisation of oestrus in cattle has proved a very effective accompaniment to AI especially in beef and dairy heifers at pasture, where thrice-daily inspection and rounding up of heifers as they come on heat is seriously impractical.

Techniques for accelerating the transmission of genetic material from superior cows include MOET (multiple ovulation with embryo transfer) and cloning (Moore and Thatcher 2006). Both of these are more technically challenging than AI, but limited by the multiplication factor. The technique of MOET may produce at maximum 60 calves from one superior cow. AI can produce 60,000 calves from one superior bull. Cloning is the process of producing identical offspring by embryo splitting or nuclear transfer. Identical twins are natural clones. Cloned cattle have been produced by the laboratory technique of nuclear transfer (as was Dolly the sheep). In theory this technique could produce large numbers of identical animals. In practice the success rate is low (the wastage rate is high). Moreover early experiments led to prolonged pregnancies in the surrogate mothers and the birth of abnormally large calves. The technique has improved but at present the financial and welfare costs grossly outweigh any real benefits in terms of genetic gain. At present the only individuals likely to benefit are individual entrepreneurs able to persuade gullible punters to pay well over the odds. There may be a place for cloning as a technique to multiply animals genetically modified to activate the production of specialist milks containing a vital pharmaceutical. This will be discussed below in the section on lactation.

Genetic engineering

Genetic engineering is the popular description for a variety of techniques for manipulating the genotype of living systems through the insertion of foreign genes into early embryos, gene deletions, or even modifications in gene sequence using gene targeting. Scientists are more likely to use the expression genetic modification, presumably because it sounds less emotive. When this technology is linked to scientific discovery of the specific genes and gene sequences associated with biological functions it creates immense possibilities to manipulate the form and function of living things in ways that we think fit. This awesome new power to do both good and harm carries awesome new responsibilities and it is right that all scientific procedures involving manipulation of the genome (other than through conventional breeding) should be subject to strict regulation at international and local level. It is, in my view, facile either to applaud or condemn genetic engineering as a matter of principle. Every planned procedure needs to be considered on its merits: independently reviewed at the outset to assess anticipated benefits and costs, and independently monitored at several stages thereafter (Maga and Murray 2010). The advantages, limitations and constraints of genetic modification of embryonic cells are discussed in the context of their potential application to aspects of animal production, e.g. reproduction, growth, lactation, disease control. It is necessary to establish *a priori* protocols for evaluation of anticipated risks and benefits attached to all planned procedures. Subsequently there must be surveillance at all stages of development to assess the extent to which the planned objectives have been met and to identify any unplanned consequences for the genetically modified animals and any other animals with which they have come into contact. Box 7.1 summarises a protocol produced by EFSA for the monitoring and assessment of risks associated with the genetic modification of farm animals.

Modifying reproductive performance, conclusions

A tremendous amount of scientific endeavour has been invested into the study of reproductive physiology in farm animals. Much of this has been good fundamental research intended primarily to increase our understanding of reproductive processes. However it would be disingenuous to deny that it has also been motivated by the desire to increase productivity through 'improved' reproductive performance. In this regard, one should not expect too much. It is often said that natural selection produces better results than the most brilliant scientist. This jibe is a little unfair: it is self-evident that natural selection (unintelligent design) produces winners, the fittest plants and animals, but there are many losers and progress is very slow. Science has the skills to speed things up. However in this branch of science, above all, it pays to exercise humility. Since reproductive success is the most important determinant of evolutionary fitness, we are unlikely to do things better. This is reflected in the fact that genetic variation, thus heritability, is low for traits linked to reproductive success, best measured in mammals by survival to weaning. We have had more success in manipulating things like growth rate, mature size, coat colour, etc., where genetic variation and heritability are relatively high, presumably because these traits have been much less important for natural selection. We may conclude that the traits that we have been most able to manipulate through selective breeding (or genetic engineering) are those of least importance to the animals themselves.

Box 7.1 Genetic modification (GM): assessment of risks to food animal health and welfare (EFSA 2012)

Potential effects of genetic modification on animal health and welfare

The intention of producing a GM animal will be to promote some form of potential benefit, e.g. increased productivity, increased feed conversion ratio, disease resistance, reduced pollution. However, while the intended effects might occur and be verified, there may be unintended side effects, e.g. an increase in disease resistance in GM individuals may lead to more silent carriers and increase the risk to others. It needs to be shown that both the intended and unintended effects do not jeopardise the health and welfare of the GM animals or other animals with which they may come into contact.

Assessment strategy

Stage A. Monitoring at the laboratory level

Objective: to define under laboratory conditions the intended effects and to determine the consequences of any possible unintended effect on the health and welfare. The procedures should include:

- Demonstration of the stability of the genetic construct.
- Health of the GM animal assessed on the basis of clinical examination and associated laboratory tests, health records, zootechnical data (e.g. growth rate, fertility and other productivity indices), relevant physiological measurements, e.g. immune function.
- Welfare of the GM animal: to include the ability of the GM animals to carry out normal behaviour, physiological functions and to develop normally.
- Welfare of the surrogate dam: to include possible effects on the surrogate dam during pregnancy, e.g. due to foetal size, endocrinological effects, including analysis of offspring aborted and born dead.

Stage B. Experimental farm assessment

Objective: to determine the health and welfare consequences on the animal of the intended and any unintended effects of the genetic modification under controlled farming conditions. The emphasis at Stage B should be directed to subtle changes in health and especially the behaviour of the GM animal in relation to other animals.

Stage C. Field trial assessment

Objective: to determine any low-frequency unintended effects in the GM animal during its use in farm settings. This should involve the surveillance

Continued...

Box 7.1 continued...

of animals on several holdings operating under normal farming practices and conditions, and should compare GM animals with comparator non-GM animals.

Post-market monitoring

The objective of post-market monitoring and surveillance is to determine unintended effects of the genetic modification on large numbers of animals and in more varied commercial conditions. This will require a mandatory procedure for reporting any adverse effects on health and welfare to be carried out by veterinarians, farmers and other authorised persons.

Growth

Some of the fundamental laws of growth were explained in Chapter 2 (Figure 2.2). For the purposes of animal production, the two most important laws are:

- Growth rate is highly correlated with mature body size: select for increased growth rate (kg/day) and you simply create bigger animals.
- As animals approach mature body size, energy requirement for maintenance increases as the animal gets bigger, and energy requirement per kg weight gain increases as the animal deposits relatively less energy as protein and more as fat (Figure 2.2).

To recapitulate: it pays to select for increased growth rates in the slaughter generation when the costs of maintaining the breeding generation are low (e.g. pigs and poultry, Table 2.1) or divergent selection can be used to produce large offspring from small mothers (e.g. suckler beef).

Body size

In this science-based chapter it is instructive to examine the laws that link body size, metabolism and maturation rate in more detail. One of the fundamental laws of energy metabolism is that metabolic rate, thus metabolizable energy requirement (ME, MJ/day) at a defined state of activity (e.g. maintenance) is related to body weight (W, kg) to the power 0.75. This is a mathematical relationship that does not really need a biological explanation. However the exponent 0.75 closely describes the ratio of body weight to surface area of the skin from which heat is lost and, more critically, surface area of the metabolically active body cells wherein the heat is generated. Thus in homeotherms:

Energy requirement (ME, MJ/day) = x. Body weight $(W)^{0.75}$

The relationship between mature size and time taken to mature can be described by a second, linked equation, namely

Time taken to mature (days) $t = y$. Mature weight $(M^{0.25})$

For practical purposes it is most convenient to measure maturation rate during the period of most rapid and reasonably linear growth from 0.25 to 0.75 mature weight. Average energy requirement for growth during this period can then be assessed at the mid-point of this range (0.5M). The relationship between the energy cost of growth, maturation rate and mature size can then be described by the following equation.

$$E = x.(0.5\ M^{0.25}).t$$

Alternatively,

$$E = (x.0.5M^{0.75})(y.M^{0.25})$$

Which simplifies to

$$E/M = y/x$$

Put into words, this means that, within a particular class of animals for which x and y are constant, the feed energy requirement for growth expressed per kg mature size is the same, i.e. size *per se* has no impact on the energetic efficiency of growth (Kirkwood and Webster 1984).

There are however some fundamental differences in the energetic efficiency of growth between different classes of homeothermic animals. Figure 7.2 plots time taken to mature against mature weight on a log:log scale for mammals and altricial (nest-fed) birds, with additional data for primates (Webster 1994). All the mammals, with the exception of primates, fall on the same straight line, which implies not only that after accounting for size all are maturing at the same rate but that all appear to be maturing as fast as they can. The class of altricial, nest-fed birds matures faster than mammals. Maturation rate in precocial birds (that have to forage for their own feed from the time of hatching) is intermediate between altricial birds and mammals. A Darwinian explanation for this is that for most animals, maturing as rapidly as possible carries an evolutionary advantage in terms of survival, which implies it has been strongly favoured by natural selection. However, for mammals and precocial birds, the evolutionary benefits of rapid growth must be balanced against the need to ensure functional mobility at all stages of development to increase the chances of survival to sexual maturity. Altricial birds do not need to be functionally mobile while confined in their nests. Thus they can take on energy at a very rapid rate early on, then

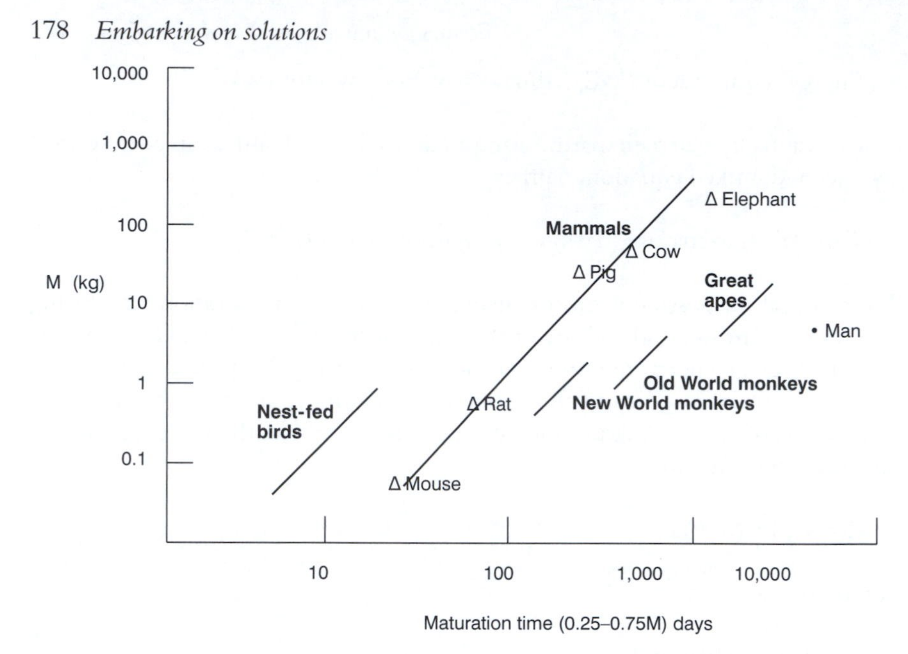

Figure 7.2 Mature body size (M) and the time taken to mature in mammals and birds (from Webster 1994). Both axes, M (kg) and time (days), are plotted on a log10

restructure themselves into perfect flying machines just before fledging, which takes place when the birds are close to mature weight.

Figure 7.2 also illustrates the evolution of a different strategy in primates: a progressive slowing in maturation rate through New-World monkeys, Old-World monkeys, the great apes to man (Webster 1994). The Darwinian interpretation here is that in primates the evolutionary advantages of slowing growth to prolong the educational advantages of childhood under supervision became progressively more conducive to fitness than maturing as fast as possible. This effect is most conspicuous in *Homo sapiens*. If we conformed to the general rule for most mammals, children would be as big as their parents by two years of age, a prospect that would terrify most mothers. Figure 7.2 provides a novel illustration of the relative importance of cognitive relative to physical ability in humans (and other primates) relative to the large majority of mammals and birds. Another logical conclusion would be that for mammals that fall on the general line, the largest animals would obtain the greatest educational benefits from prolonged childhood, like elephants – for which the evidence is rather good (Nissani 2004).

The fact that mammals and precocial birds have evolved to ensure fitness through functional mobility at all stages of development places constraints of the ability of man to manipulate either growth rate or shape in the animals farmed on land. Fish, swimming weightless in the sea, are more open to manipulation. Farmed transgenic salmon produced by microinjection of a DNA construct consisting of growth hormone sequences consume 2.5 times as much feed and grow at rate 2.5 times the rate of conventional salmon (Cook *et al*. 1995). Gross feed conversion efficiency is improved by 10 per cent. This can be attributed simply to higher feed intake relative to maintenance (see Figure 2.2). There appear to have been no major locomotor or welfare problems for these animals when reared in confinement, presumably because of the minimal problems of supporting body weight in an aquatic environment. There are potentially serious but largely unknown environmental issues if these animals are allowed to escape. In confinement they outcompete smaller fish. However, in nature, their exaggerated size may be associated with reduced fitness. Since we don't know, we should apply the precautionary principle and produce sterile genetic constructs unable to survive one generation.

Manipulation of growth rate and body composition in terrestrial mammals and birds (e.g. chickens and turkeys) is inherently more likely to cause problems, especially locomotor problems, in animals where muscle development and body weight get too far ahead of skeletal development. The locomotor and cardiovascular problems of fast-growing strains of broiler chickens have been described already (pp. 146–7). In a Darwinian sense, the intensive broiler industry is founded on strains of birds that are lethal recessives. Fed *ad libitum,* a high proportion would not survive to reach breeding age.

Several groups have succeeded (up to a point) in accelerating growth rate in mice, rabbits and pigs through transgenic insertion of a foreign gene that expresses growth hormone (GH). Early trials were fairly catastrophic. The notorious Beltsville pig was crippled by failure of skeletal development to keep pace with weight gain. The welfare problems have been reduced by refinement of the technique to prevent over-expression of GH. However, in reviewing progress Pursel *et al.* (1990) reported (inter alia) that

> the percentage of injected ova that developed into transgenic pigs varied from 0.31 per cent to 1.73 per cent. The percentage of transgenic pigs that expressed the fusion gene ranged from 17 per cent to 100 per cent. Elevation of growth hormone (GH) in expressing transgenic pigs improved feed efficiency about 15 per cent, and markedly reduced subcutaneous fat compared to non-transgenic siblings. Growth rate was enhanced in some transgenic GH pigs but not in others, possibly due to dietary limits. The 'over-expression' of GH was detrimental to the general health of most transgenic pigs. The most prevalent problems were lethargy, lameness, and gastric ulcers. Gilts that expressed foreign GH genes were anoestrous. Boars that expressed foreign GH genes lacked libido.

Given this evidence one hardly needs to recruit ethics to conclude that this exercise is a commercial non-starter.

Body shape

We have, through selective breeding, had considerable impact on the shape and body composition of domestic animals, mainly with the intention of making them more meaty by increasing the ratio of muscle to bone and other non-saleable products. Up to a point this has been achieved without compromise to fitness, in a domesticated environment, or animal welfare. The most extreme departures from normal body shape are seen in the enormous expansion of the breast muscle in broiler fowl and, especially, turkeys. This has been possible because these are essentially non-functional muscles in birds that don't fly. The expansion of the breast muscle in the turkey has reached the point where the males cannot mate normally (they fall over backwards), so all breeding has to be by artificial insemination. This would, of course, be a lethal development in the wild. There is not, to my knowledge any good evidence to indicate that this presents significantly greater welfare problems for these turkeys than the artificial breeding techniques used for other farmed animals.

One genetically generated distortion of body shape that does present welfare problems arises from the selection for extreme muscularity in beef cattle. The Belgian Blue Cattle breed is notorious for its exceptional muscular development known as 'double-muscling'. This extreme phenotype is due in part to a loss-of-function deletion in the myostatin gene that, as its name implies, restricts further muscle development when enough is enough. This gene deletion seriously compromises fitness since the abnormal muscle development is accompanied by a reduction in the dimensions of the pelvic cavity, which greatly increases the risk that natural parturition will be difficult or impossible. Double muscling does occur spontaneously in other heavy beef breeds (e.g. Charolais, Piemontese) but since the cows are expected to live in a natural environment with minimal interference, the heifer calves are not retained for breeding. In the Belgian Blue however there has been a deliberate policy to fix the gene within the breed. Cows carrying the trait are housed before calving and many submitted to repeated premeditated Caesarian section. The animals have other phenotypic abnormalities including distortion of the upper respiratory tract that renders them more susceptible to respiratory infections and distress during exposure to heat or exercise. The high risk of both health and welfare problems for truly double-muscled beef cattle means that even the most heartless producer would not consider rearing them in large numbers for slaughter. The role of this genetic freak is to produce a small number of bulls whose semen can be used for insemination of conventional cows, typically Holsteins, to produce a heavily muscled calf for beef production from the dairy herd. These cross-bred calves fetch good prices and present no special health and welfare problems during rearing. Moreover obstetric problems are no higher than following insemination with other large beef breeds (e.g. Charolais) because the pelvic dimensions of the Holstein mother cows are normal. The ethical question for the dairy farmer is whether or not to select semen from bulls from a breeding programme that is known to be an abuse to welfare.

Lactation

The two obvious challenges presented to animal scientists are to increase milk yield and increase production. The increase in lactation yield of dairy cows through a combination of improved nutrition, genetic selection and rapid dissemination of high-merit genotypes, has been striking; a doubling within the last 20 years. The details of this progress have been described earlier. One point that bears repetition here is the current trend to adjust selection indices for dairy cows to give less emphasis to yield in first lactation and more to traits associated with robustness and lifetime performance.

One of the most important hormones involved in the natural synthesis of milk is growth hormone (GH) or Somatotropin. In 1994 Monsanto licensed a synthetic form of bovine GH using recombinant DNA technology to create recombinant bovine somatotropin (rBST). Early studies showed that regular injections of rBST to dairy cows, starting at about day 50 of lactation increased total lactation yield by 11–15 per cent. Early suggestions were that this might become an offer that dairy farmers could not afford to refuse. Inevitably it aroused much public concern. The use of rBST has been prohibited within the European Community and it is now, in effect, only being used routinely within USA. Some of the original objections to rBST were not strongly supported by scientific evidence. However more recent meta-analysis reveals that the 11–15 per cent increase in milk yield has been accompanied by a 25 per cent increase in clinical mastitis, a 40 per cent reduction in fertility and 55 per cent increase in lameness (Dohoo *et al.* 2003), evidence that the effect of rBST was to push the cows even further beyond the limits breached by selection for yield at the expense of robustness. Currently the use of rBST in the USA is less than 15 per cent. It was not a good idea, and on the present evidence no case can be made for future investment in high science with the simple aim of increasing milk yield *per se.*

There is little point in seeking to use high science in order to manipulate the composition of milk as synthesised by the cow (goat, ewe, buffalo, reindeer or yak) since there are so many nutritious and delicious dairy products that you can create after the milk leaves the animal. There is however some potential for the genetic modification of dairy animals to produce valuable specialist proteins in milk. The words 'Pharming' has been coined to describe the insertion of DNA containing foreign genes into sheep, cows and goats to express in their milk pharmaceutical proteins such as blood clotting factors or alpha-1-antitrypsin for the treatment of emphysema and cystic fibrosis. Another intriguing exercise has involved the expression of the extremely strong spider's web protein into the milk of goats. In all these cases the success rate tends to be very low, both for the initial genetic modification and for the multiplication of the genetic construct by cloning. Whether or not any of these exercises in genetic engineering will become a commercial proposition will be governed by three main measures of cost and benefit.

- What is the potential 'use-value' of the product?
- Can it be produced more easily/cheaply by other means?
- Will the public accept it?

Many complex proteins used in medicine (e.g. insulin) can be synthesised by bacteria. In some cases, e.g. human alpha-1-antitrypsin, the protein needs to be activated by a process of glycosylation, which only occurs when it is synthesised in mammals. In these circumstances, synthesis of vital proteins at relatively low cost in the milk of sheep can make the difference between life and death for adults and children with genetic deficiency disorders. Attitudes of individuals and society to the genetic engineering of animals are driven by a number of complex emotions: concerns for animal welfare, concerns about scientists playing God, fear of the unknown. Some will undoubtedly conclude on ethical grounds that the genetic modification (GM) of animals is an unacceptable practice in any circumstances. Others will consider each case on the basis of a more or less informed cost–benefit analysis. Each proposed experiment in GM should be independently examined at the outset and reviewed at each stage of development as outlined in Box 7.1. This approach should have concluded that the genetic engineering of a pig to produce doubtful economic gain at serious cost to the health and welfare of the animals concerned should be prohibited at Stage B. In the case of rBST injections for dairy cows, it is fair to conclude that the real (and anticipated) costs could not be proven until the post-marketing stage. However, in both cases, I believe there is enough evidence to justify a ban.

The technique for insertion of the gene to express alpha-1-antitrypsin in sheep's milk involves a lot of wastage through failure to express the gene, failure to carry the modified embryos to term, failures in cloning. The development stage undoubtedly carries some welfare costs to the animals, which are taken into account in the regulation of scientific procedures under the UK Animals (Scientific Procedures) Act 1986. However, once established, the health and welfare of the animals appears to be entirely normal. Indeed, because they are so valuable, they are likely to be treated with much more individual care than ewes on extensive sheep farms. The benefit is not just the production of a little more of one sort of food within an infinite choice range, but a unique drug that can keep children alive. In my opinion, the ethical case for this sort of GM is overwhelming.

Health

At this stage you may have formed the impression that there is little more animal science that needs to be done to improve the productivity of farm animals through the sciences examined so far, i.e. nutrition, physiology and genetics and, on the whole, I think you would be right. However when we consider the matter of animal health, we shall never reach that stage. New diseases continue to emerge, partly because infectious organisms (viruses, bacteria and parasites) mutate faster than farm animals, and partly because climate change and international movements of humans and animals create increasing risks of disease transmission. Moreover, high-output production systems combine to increase the challenges to animal health from infectious organisms and physiological stresses, while tending to reduce the ability of the animals to cope with these challenges and stresses.

As discussed in Chapter 3, the most effective approaches to the promotion of good health in farm animals involve the application of conventional wisdom in matters of good husbandry. This includes good feeding, good housing, good hygiene, biosecurity, the appropriate application of preventive medicine (e.g. use of vaccines) and the early diagnosis and treatment of diseases and disorders that may arise despite these precautions. The commercial farmer should respect the health and welfare of each individual animal, but for reasons of economics and common sense the main aim must be to preserve the health of the flock or herd. Science can continue to contribute to this aim in the laboratory through the development of new vaccines and chemotherapeutics and through the discovery and development (by several means) of animals that have greater or absolute genetic resistance to specific diseases, especially epidemic infections associated with a single specific pathogen. In practice, however, many of the most intractable health problems for farm animals are the complex, multifactorial endemic and production diseases (e.g. post-weaning enteritis in pigs, lameness in dairy cows) that cannot be attributed to a single cause or resolved by a single injection. Most competent farmers and veterinarians have a reasonable understanding of the major risk factors for these conditions and use this knowledge as the basis for planned herd health schemes. However good systems science, including the meta-analysis of records from multiple herd records, can do much to refine this approach by quantifying the relative importance of presumed health risks, identifying unforeseen risks and prioritising actions for health control according to HACCP (hazard analysis and critical control point) principles (Gardner 1997).

The case for the continued development of new and improved vaccines to protect farm animals from infectious diseases is so obvious that it needs no advocacy. Unfortunately a good vaccine that needs to be administered only once a year or once in a lifetime to give lasting protection generates less income for the pharmaceutical industry than a medicine that needs to be taken every day. Vaccination makes good commercial sense in the intensive poultry industry where the number of individual birds is colossal and their life span is short. World egg production is currently about 60 million tonnes per annum. Some basic arithmetic reveals that the international market for a vaccine for each of the major infectious diseases (e.g. Newcastle disease, avian influenza) is about 2 billion (2×10^9) per annum. There is much less commercial incentive to develop vaccines for localised diseases in low-value animals in poor countries.

There is no doubt however that the development of vaccines and the use of international aid to ensure the effective implementation of vaccination programmes has made, and can continue to make, a massive contribution to the success of animal farming, particularly in the third world and in pastoral communities where dependence on livestock is essential to human survival. Rinderpest, the great plague of cattle, was almost eradicated by the 1970s. Failure to sustain the control programme through periods of political turmoil in Africa caused a temporary resurgence of the disease but now we can conclude with some confidence that the disease has been eradicated from the earth, like smallpox in humans. There are many other programmes to control or preferably eradicate important infectious diseases of animals such as brucellosis, a zoonotic disease of grazing animals such as cattle and goats that can cause severe chronic disease in humans. It is unrealistic to expect major support from the pharmaceutical companies for the basic science and implementation necessary to improve the health and productivity of the animals on whom these poor and pastoral communities depend. This role has been taken on by the international aid agencies and it can yield great rewards to all parties. The old Oxfam adage 'Give a man a fish and you feed him for a day, teach him to fish and you feed him for life' may be extended to read 'help him (and especially her) to keep the fish/hens/goats/cattle fit and well for the whole of their productive lives and s/he will not only have food for life, s/he should have some income too, and with this, some dignity'.

The most promising new approach to the control of infectious diseases in farm animals is through identification of genetic variability in resistance to disease, in particular genetic variation in the non-specific and specific elements of the immune response. A great deal is known about genetic variation in host resistance or tolerance among the major domestic livestock species. Well-known examples include Marek's disease in chickens, F4 and F18 *Escherichia coli* infections in pigs, and nematode infections, mastitis, dermatophilosis, trypanosomosis and theileriosis in ruminants (Gibson and Bishop 2005; Soller and Andersen 1998; Hawken *et al.* 1998). There are breeding programmes that aim to select animals for enhanced resistance (or tolerance) to these diseases. This approach is particularly attractive in the context of infectious diseases of farm animals in the third world where farmers are seldom willing to control disease through an eradication policy or able to exercise strict biosecurity. Moreover they are seldom able to meet the costs of vaccines and chemotherapeutics. Resistance or tolerance to pathogens (whether viruses, bacteria, protozoa or nematodes) can be increased through identification of naturally resistant individuals and breeding lines. Unfortunately the success of this approach depends on very accurate records of disease in individuals or inbred lines. In practice, this is likely to limit it to large high-tech operations for poultry, pigs and dairy cattle. It is much less likely to succeed in third-world situations with many small farms, and little or no hard evidence of breeding or disease incidence.

There are other constraints on breeding for disease resistance. Most of the infectious diseases of importance involve more than one gene or linked combination of genes. Differences in the expression and severity of clinical disease are not simple functions of genetic differences in resistance but can involve environmental factors and variation in immune responses associated with concurrent stressors. In consequence, improvement to disease resistance through conventional genetic selection is likely to be very slow, especially in species like cattle with low reproductive rates.

The new science of genomics and techniques for gene manipulation offer great potential for accelerating the rate of genetic improvement in resistance to infectious disease. They have not yet had a major impact on farm animal health and disease control but their potential is enormous. In genomic approaches to the improvement of disease resistance, the criterion for selection is shifted from phenotypically expressed disease status to allele status at the DNA level. This mode of selection is termed 'marker-assisted selection' or MAS. In principle, MAS can enable selection for disease resistance without exposure to disease challenge and allows highly accurate selection which is unaffected by environmental factors. The procedure involves the following steps (Soller and Andersen 1998):

- construction of genomic maps for the species of concern;
- genomic analysis of the resistance trait, leading to mapping and identification of potential disease-resistant loci (DRL);
- association of genetic differences within the putative DRL and clinical information on breed and strain differences in resistance to specific infectious diseases;
- incorporation of information on DRL into MAS-based genetic selection for increased disease resistance;
- dissemination of the resistant strains by conventional breeding.

The details of all this are far beyond the scope of this chapter. At first glance it may appear too complex and expensive to be practical, particularly for diseases of tropical livestock where the returns on investment are unlikely to be substantial. However the science of genomics is well advanced and the obstacles are not too severe. Workers in this area of veterinary genomics have been given a head start by the fact that there are many similarities between the genomes and the location of the DRL in farm animals like chickens, pigs and cattle and those in the extensively researched mice and men. As an example one can identify genetic variation as a DRL in *Bos taurus* and the N'Dama cattle (*Bos indicus)* that are known to be resistant to trypanosomiasis. Having identified the DRL in N'Dama cattle one can then look for it in other cattle, possibly more productive breeds, and use the information as the basis for MAS.

In the USA geneticists have identified dairy cow genotypes with increased resistance to mastitis caused by *Staph. aureus* (Wall *et al.* 2005). This is scientifically exciting but may not have a great deal of impact on the dairy industry since *Staph. aureus* is but one of many bacteria that can cause mastitis, and one that can be controlled fairly effectively by conventional means. More promising news comes from researchers in Ames, Iowa, USA (Casas *et al.* 2011) who have been studying resistance to multiple diseases in beef cattle. A genetic marker associated with resistance to multifactorial respiratory diseases of feedlot cattle was found on bovine chromosome 20. This genetic marker is in very close proximity of several markers related to other diseases, such as Johne's disease and bovine viral diarrhoea. The implication is that this particular region on chromosome 20 may have a significant effect on general health in animals, not just cattle. If this is confirmed it presents a powerful tool for a marker-assisted selection programme that could radically increase the resistance of farm animals to endemic diseases associated with a wide range of pathogens. However one should not get too excited too soon. If it is true then I must assume that it involves some elements of the non-specific immune response, e.g. cytokine mediation of inflammation. It will need immunologists to discover just what is different about these animals.

It is difficult to raise ethical objections to the application of genomics to the control of infectious disease in farm animals since the high science is done in the laboratory with minimal interference to the animals and the direct aim is to improve animal health. Obviously the ultimate benefit to us comes from increased productivity and profit. However improving productivity through increased resistance to infectious disease has to be good for them too.

Before the decision is taken to embark on an expensive programme of disease control through modern genomics, it is necessary to address certain epidemiological questions in order to decide whether or not it is the best approach. It is necessary to establish whether the genetic variability is linked to disease *resistance*, which implies rejection of a pathogen, or *tolerance,* which implies acquisition of infection but with little or no clinical symptoms. Increased resistance is good for the exposed individual and the population, since it radically reduces the capacity of the pathogen to spread. Increased tolerance can increase transmission of infection, thereby increasing the risk of clinical infections in more susceptible animals. Moreover an increase in healthy carriers greatly increases the difficulties associated with introducing a programme of eradication. This problem has particular significance in relation to eradication of the prion diseases like BSE (bovine spongiform encephalopathy) and scrapie in sheep. There are clear genetic differences in susceptibility, as measured by the onset of clinical disease or the appearance of prions in the central nervous system, but these do not guarantee that the apparently resistant animals are free of prions. One may conclude that the genetic approach is most appropriate for those diseases where eradication at the farm, national or international level, is not an option and vaccination has met with limited success.

Animal welfare

While most farm animal science has inevitably been directed towards increasing productivity, the main aim of the emerging discipline of animal welfare science has been remedial, much of it intended to identify and remedy systematic failures of intensive systems to meet the physiological and behavioural needs of the animals. In previous Chapters 3, 5 and 6 I have described protocols for the characterisation of animal welfare (e.g. the Five Freedoms), the assessment and surveillance of farm animal welfare and the promotion of welfare standards that are satisfactory (or better) through legislation or competitive quality assurance schemes within the privates sector. One essential role of animal welfare scientists is to provide justification, based on evidence, for the procedures adopted to assess the welfare of animals on farm largely from animal-based measures. The pan-European Welfare Quality® project has produced excellent, science-based, protocols for the on-farm welfare assessment of cattle, pigs and poultry (Welfare Quality 2009).

There will always be some need for new research into farm animal behaviour and the extent to which this behaviour is compromised, distorted or abused on farms, in transport or at the point of slaughter. There is a real danger however that government-sponsored research of this kind can hinder, rather than advance, progress towards improved standards. Governments routinely state that their animal welfare policies are based strictly on science, even when, as discussed earlier in the example of pregnancy stalls for sows, different governments have reviewed the same science and come to opposing value judgements (p. 105). The desire of politicians to protect themselves from the public by taking refuge behind the scientists can lead to the commissioning of much trivial research designed to confer scientific credibility on what would be obvious to any good stockman. It can also lead to an unholy alliance between scientists and politicians to delay decisions that would clearly benefit the animals pending (to use the most egregious of scientific clichés) 'considerable further research in this area'.

There are some excellent opportunities for interesting and important animal welfare science. One is motivation analysis, the study of what drives sentient animals to do what they do, what they fear and what they favour. Much of this can be achieved using current techniques in ethology and experimental psychology, including the use of pharmacological agents with known actions on behaviour and mood (affect). In my view, though I may at some stage have to recant, we can discover most of what we really need to know about how sentient animals feel and think without recourse to new techniques in neurobiology, e.g. mapping brain pathways within the CNS or identifying the sites of action of neurotransmitters. Such research can have significant benefits for human medicine and the pharmaceutical industry but, at present, I do not see how it can increase our understanding of animal welfare. In any applied science it is important not to be seduced by 'gee-whizzery'. Every 'Gee Whiz!' should be accompanied by a 'So what?' If the answer to 'so what?' is likely to be 'not much' then better to forget it.

Conclusions: Who needs animal scientists?

The central theme of this book is an exploration of the contribution of farm animals to 'planet husbandry'. Variations on this theme have embraced (inter alia) animal biology, veterinary medicine, environmental physics, economics and ethics. During my career as a scientist mostly involved in the study of farm animals, I have never doubted its importance but become increasingly aware of its limitations. Farm animal science is limited in scope because it relates only to those animals used by man in agriculture and, as I have explained in this and previous chapters, the potential to manipulate animals to improve production is limited and likely to compromise fitness. Given these constraints, how best can animal scientists serve the society of humans and other animals?

It is terribly easy to mock scientists. Science itself may be disinterested and rational but we too are sentient creatures, powerfully motivated by the desire to feel good about ourselves. We seek the respect of our peers and the tangible expression of that respect in the form of research grants and research ratings. From the moment we embark upon our PhDs we are directed by our peers (though not society) to 'focus'. Three years into a post-doctoral fellowship and our focus is likely to have progressed to a severe case of tunnel vision. I accept that the greatest of new scientific discoveries are likely to be made in the laboratory. However the greatest increases in understanding tend to emerge from those with a broader vision. Darwin developed his vision on the basis of deep thought about simply everything.

I propose therefore that what animal scientists need is less emphasis on focus and more on panoramic vision. Reductionist science is essential for the business of new discovery. However the greatest need for animal science is not for new discovery but for better understanding of how animals, not just farm animals, may best fit into sustainable production within the living environment. I include the word production because the living environment must produce things of value to us otherwise we can't afford to sustain it. I shall be discussing these things in more detail in the final two chapters. At this stage I suggest that the animal scientist should be competent to understand and, where relevant, practise the specialist reductionist disciplines that underpin the scientific understanding of farm animals. However we shall be much more use if we can also interpret and integrate these underpinning sciences in ways that can be understood by the policy-makers and the public at large. In this way we can help to ensure that science makes its proper contribution to action for the production of better food (etc.) from fitter and happier farm animals in a healthy, sustainable and beautiful environment.

8 Futures for farm animals in planet husbandry

> The care of the Earth is our most ancient and most worthy, and after all our most pleasing responsibility.
>
> (Wendell Berry)

The quotation that prefaces this chapter comes from Wendell Berry, poet, philosopher and all-round friend of the earth. According to him, the good life 'includes sustainable agriculture, appropriate technologies, healthy rural communities, connection to place, the pleasures of good food, husbandry, good work, local economics, the miracle of life, fidelity, frugality, reverence, and the interconnectedness of life' (see Wendell Berry, in wikipedia.org). It should be self-evident that I am a disciple. The central aim of my book is to present a constructive case for the future of farm animals in sustainable agriculture, a case moreover where the advocacy is supported by sound science and hard evidence. Sustainable agriculture is a fashionable and reasonably well-understood phrase and I used it early on to give readers a feel as to what the book is all about. However, as I also wrote at the outset, my intention has been to broaden the discussion beyond the classical definition of agriculture as 'cultivating fields' to incorporate all major issues relating to the proper management of land and all that live off the land. The three principles of planet husbandry should be:

- efficiency in the use of resources for production of goods;
- humanity in the management of farm animals and wildlife;
- sustainability in the stewardship of the living environment.

The resources of the land must be managed efficiently and sympathetically to create goods and services for immediate use, ranging from food to footpaths. The farm animals who contribute to the husbandry of the land and the feeding of the people should be managed in such a way as to achieve a fair balance between efficiency in the use of resources, and humanity based on respect both for their sentience and their contribution to the enterprise. At the same time the value of the land must be preserved and enriched as assessed by measures of sustained production, conservation of natural resources, climate control, viable rural communities and beauty itself. All three aims, humanity, efficiency and stewardship, are embraced within the single word husbandry. In Chapter 1 I dropped the comfortable and clichéd phrase sustainable agriculture in favour of the more ambitious (and more accurate) phrase 'planet husbandry', defined as follows:

> The aim of 'planet husbandry' is to manage the land in such a way as to produce goods and services of value and to enhance the quality of life for humans, animals (domestic and wild) and all elements of the living environment. This aim should embrace not only our immediate needs, it must also be directed towards the sustainable management of the resources of the planet.

Aims and opportunities

Aims and opportunities for total agriculture, now called planet husbandry, were identified in Table 1.2 as production of food and other goods for human use, management of the land as an amenity and stewardship of the environment. Every opportunity for the individual carries a matching responsibility. These responsibilities do not relate simply to the provision of food and other goods for humans; they extend to sentient farm animals and, indeed, to the welfare of all life (flora and fauna) on the land. Moreover the responsibilities apply not just to the farmers and landowners but to us all, because we all live off the land. A comprehensive review of all the major contributors to planet husbandry would be an enormous undertaking and well beyond my pay grade. The task I have set myself is to review the role of farm animals in planet husbandry. Animal farming is under pressure from many sides on grounds of diet and human health, animal welfare, overconsumption of valuable resources, social inequalities, pollution and global warming. The validity of these various criticisms is, of course, open to debate and I have done my best to address some of them without prejudice. What is certain is that, at the planet level, the expansion of livestock farming and consumption of food from livestock has to slow down and eventually establish a sustainable limit. Those who take the vegan or vegetarian options, or simply eat less meat, can make a significant contribution to the reduction of total livestock production and the associated environmental costs. However animal farming won't just quietly fade away, not least because it can't. Pastoral systems occupy 26 per cent of ice-free land on the planet. A further 33 per cent of cropped land goes to animal feed. Thus directly or indirectly well in excess of 50 per cent of managed land is devoted to animal agriculture (FAO 2006). Moreover, despite all the above concerns, world-wide consumption of meat, milk and eggs is predicted to rise by about 70 per cent in the next ten years. There is no alternative to animal husbandry. We just have to do it better.

At present, we are not doing it very well. According to the FAO report *Livestock's Long Shadow* (2006) 'the livestock sector is the major driver of deforestation, as well as one of the leading drivers of land degradation, deforestation, pollution, climate change and facilitation of invasion by alien species'. Extensive pastoral systems are a major cause of land and soil degradation, including loss of sequestered carbon. Mass production of feed crops (e.g. soya beans) for intensive livestock systems has been a major contributor to deforestation; intensive housing for livestock has been a major contribution to pollution. FAO's concern about 'invasion by alien species' sounds especially apocalyptic but I guess they are not referring to extraterrestrials.

One of the key arguments in Chapter 1, 'Whatever happened to husbandry?', was that, for 99 per cent of the history of agriculture, farm animals seldom competed with humans for resources. They were part of a sustainable system, subsisting largely on complementary feeds and providing goods, services and income for those directly involved in their management. The main driving factors for industrialisation of livestock farming in the 20th century were machinery, cheap energy from fossil fuels, and antibiotics. However other socioeconomic factors were in play. Mass migration from the country to the cities, in both rich and poor countries, driven by the hope for more money and better living standards, created an ever-increasing market for mass-produced food. At the same time the large and increasing disparities of wealth between the rich and the poor increased the opportunities for farmers to utilise 33 per cent of cropped land to grow feeds for animals, to feed those that could afford it on meat, eggs, beer and multiple cheeses. Much, not all, of this can be classed as competitive food production. Not all, because very large quantities of crop residues from cereals and oilseeds grown for human consumption are incorporated into animal feeds, e.g. brewer's grains, maize gluten, soya bean meal, although in the final example the high-protein crop 'residue' is arguably more valuable than the oil.

It is fundamental to human behaviour that most people who can afford to eat meat and dairy products choose to do so. In the last 50 years, per capita consumption of food from animals in East Asia has risen fourfold, from about 100 to 400 kcal (400 to 1,200 kJ) per day (Speedy 2003). We citizens of the affluent West are however in no position to criticise, since our average daily intake is about twice that (800 kcal/day). People in sub-Saharan Africa are still constrained to about 100kcal/day by poverty, not by choice. I am tempted, but resist the temptation, to launch into a polemic about man's inhumanity to man and the mindless (rather than wilful) unfairness of it all. However I shall stick to my brief, which is to review the science and humanity that underpin the proper care of farm animals and the living environment, i.e. the business of good husbandry.

My exploration of the future for farm animals in the context of planet husbandry concentrates on the main options, namely grasslands and agro-forestry, high-input industrial systems, low-input village and peri-urban systems, including 'middle class peasantry', the movement back to the land within the relatively affluent in developed societies. Each option is examined according to the three core principles of good husbandry: efficiency, humanity and stewardship. These core principles should be seen as complementary, rather than mutually exclusive. Every system from the most intensive poultry unit to the most extensive sheep farm has the responsibility to address all three principles and the right to receive reward in proportion to their contribution to the overall value of the enterprise.

Grasslands and agro-forestry

Grasslands, including rangelands, shrub lands, pasture lands, park lands and croplands sown with pasture, trees and fodder crops, represent 70 per cent of the world's agricultural area. The soils under grasslands contain about 20 per cent of the world's soil carbon stocks but these stocks are at risk from land degradation. The Land Degradation Assessment in Drylands (LADA) recently estimated that 16 per cent of rangelands are currently undergoing degradation. They state: 'Overgrazing is not only a major contributor to land degradation and deforestation, it is also responsible for substantial methane and nitrous oxide emissions from ruminant digestion. The combination of land degradation and methane emissions constitutes a major threat to climate stability' (LADA 2006). This criticism is valid but refers to unsustainably managed grasslands grazed by domestic animals, mostly the property of herdsmen for whom the immediate needs to support self and family far outweigh their long-term commitment to the environment. This is not to condemn herdsmen, be they poor Siberian nomads or relatively rich American cowboys; just an acknowledgement of the simple economics of human behaviour. We are all motivated to act by the expectation of reward. So long as action designed to sustain and enrich the environment receives little or no reward, people won't do it because they can't afford to.

Until man acquired dominion, grasslands grazed by herbivores were highly successful, sustainable ecosystems. The prairies of North America (largely treeless) grazed by buffalo and the savannah of sub-Saharan Africa (mixed grasses and shrubs) grazed by multiple species of herbivores were, in the long term, ecologically stable. In climates where droughts and bush fires occurred every year, and more sustained droughts every few years, the grasslands outcompeted the forests. In terms of planet husbandry, grass has undoubtedly been a success. The phenotypes and the numbers of animals that both grazed and fertilized the grasses were determined by the grass supply, and the phenotypes and numbers of predatory carnivores were determined by the herbivore supply. In this Darwinian cycle of nature, it was the grass itself that dictated population numbers of animals all the way up the food chain from herbivores to carnivores to human hunters, such as the North American Indians. In conditions of prolonged drought, thousands of herbivores would die to the advantage (for a time) of the carnivores and scavengers. In normal years, however, control of the animal population was less Malthusian. Ruminants and other grazing animals are extremely well adapted to withstand long periods of undernutrition during the dry season in the tropics or winters in temperate and boreal zones (see p. 80). During this time it is natural for them to lose substantial amounts of weight and body condition. The response of female animals to loss of body

a b

Figure 8.1 Sustainable and unsustainable grasslands: a) North American bison on the open prairie (© Jim Parkin, shutterstock); b) goats in Africa (© posztos (colorlab.hu), shutterstock)

condition is to stop ovulating, thereby postponing reproduction until sufficient feed is available for both mother and offspring. This mechanism helps to sustain optimal population numbers and therefore the long-term survival of the species at a relatively low cost in terms of suffering, starvation and death. While the grass above ground can appear denuded in times of drought or after a fire, it will always recover so long as the roots are unharmed. In this regard natural selection favours ungulates without upper incisor teeth that are less able to pull up the roots.

In the wild state therefore, the first constraint on animal population numbers is the amount of grass and browse available during the period of dormancy (dry season or winter). This constraint means that there is likely to be sufficient feed to support the population during the growing season and little long-term risk of overgrazing. As with all natural cycles a long-term equilibrium is achieved between flora and fauna. Disturb the equilibrium by adding *or taking away* animals (both herbivores and their predators), and the productivity and sustainability of the plant communities will be compromised. Degradation of grasslands did not occur until man interfered to distort the natural balance by seeking to draw more from the ecosystem than it could sustain. Paradoxically one of the biggest threats to the sustainability of grasslands is the practice of providing supplementary feed for the animals during the non-growing season because it encourages the herders to retain more animals, thereby increasing grazing pressure during the growing season. In areas of the UK that retain commoners' grazing rights, such as the New Forest, the feeding of hay to ponies kept out over winter is prohibited. By the end of a hard winter a number of the ponies can be very thin, which upsets a number of animal welfarists. However, in an ecological sense, the prohibition works very well.

Sustainable nomadic communities, such as the reindeer herders of Lapland, do husband the resource of the land for the long term within a harsh but relatively predictable climate. Traditional sheep and goat herders around the Mediterranean and Middle East have also timed many of their religious festivals involving the slaughter of large numbers of lambs and kids to coincide with the end of the rainy, grass growing season, leaving a smaller breeding flock to carry through the dry season. In northern latitudes we slaughter lambs when the grass stops growing at the end of the summer. However we retain a cultural/religious demand for lamb at Easter time, i.e. the end of our winter, a custom that has been well exploited by New Zealand sheep farmers.

Most of the severe problems of grassland degradation have occurred in the tropics and sub-tropics. Some of this can arise from simple bad management, primarily overstocking, and it is tempting to attribute this to poverty and ignorance. However, I repeat, one should hesitate before condemning the African herder who appears to our eyes to keep too many cattle. The risk of prolonged and unpredictable droughts is far greater in the tropical grasslands than in temperate and boreal zones. Tropical droughts bring poverty and desolation. Animals that can be killed and eaten or sold, albeit at a loss, can keep families alive for longer than crops that fail altogether.

To summarise the argument so far: permanent unmanaged grasslands and parklands (mixtures of grasses and deciduous trees), grazed by herbivores that are themselves harvested by carnivores (including humans), have through natural selection evolved to become the dominant stable ecosystem over more than 50 per cent of the ice-free, green planet. The grass needs the animals and vice versa. Thus, contrary to much current paranoia, ruminants and other herbivores grazing grasslands do not necessarily constitute a threat to our joint survival through land degradation and global warming. Animal farming is only a threat to the environment when it is allowed or forced to disrupt the stability of the ecosystem through economic pressure. In this case the economic pressure arises from the fact that the income generated from the sale of meat and milk from animals drawing all their sustenance from grasslands is insufficient to sustain quality of life for the land and those who work the land. Husbandry of the grasslands presents probably the most extreme illustration of my big theme. The value of the land for all that live off the land is too great to be sustained simply from the sale of food as a commodity.

Well-managed grassland and agro-forestry systems can contribute positively to planet husbandry in the following ways:

- highly nutritious food for humans from plants we cannot digest;
- carbon sequestration;
- water management and flood control;
- countryside amenities for human recreation;
- sustainable habitats for wildlife;
- space for windmills.

Most or all of these aims can be pursued simultaneously since there is little or no conflict of interests. The relative contribution of the different elements to total value will depend on local conditions of topography and climate. Intensively managed high-producing grassland will generate most wealth from food production, especially dairy products, but can still act as carbon sinks (Table 2.5). Extensive grasslands in the mountains and moorlands can contribute a great deal to water management, countryside amenities, wildlife habitat and energy generation from wind power. In the UK they are also the principal source of breeding stock for the stratified sheep industry. Their overall contribution to the national economy should be rewarded in keeping with the net value of each of these elements. I shall discuss how this might be done in the final chapter.

It is not possible, at this time, to reach any firm conclusions as to the overall impact of grasslands and parklands grazed and browsed by ruminants on carbon balance and contribution to/mitigation of greenhouse gas emissions. What is certain is that C balance can vary widely from high net release to high net capture according to circumstances. Here I can, without shame, fall back on a cliché. This subject really does need considerable further research. One avenue of enquiry that I find particularly fascinating is the role of silica leached (especially) from grasslands in the maintenance of colonies of *diatoms*, a major group of algae that are one of the most common types of phytoplankton (Packer 2009). A characteristic feature of diatom cells is that they are encased within a unique cell wall made of silica called a frustule, and it would appear from recent research that silica is usually the first-limiting factor for the development of colonies of diatoms (Treguer and Pondavin 2000). Diatoms in marine and freshwater environments capture solar energy and carbon dioxide by photosynthesis to create organic matter. Not only are they a primary feed source for marine organisms, it has been calculated that diatoms that sink to the sea-bed may be responsible for 25 per cent of total sequestration of atmospheric carbon dioxide. Silica originates from the earth so substantial quantities of silica are released in variable amounts into rivers from bank and soil erosion. However silica is also taken up in large quantities by grasses. The ungulates eat the grasses, and excrete the silica, which is carried into water-courses on a regular basis. By this measure it would appear that the positive contribution of grasslands and the animals that graze them to carbon sequestration can extend far beyond their boundaries and out to sea.

Conservation grazing

This is a well-recognised practice for the management of wilderness areas, nature reserves and sites of special scientific interest. The practice is based on the principle that the fitness of plant species depends as much on the animals as vice versa. Areas that are very large and relatively untouched by human hand, such as the American prairies before the arrival of the settlers, can be relied on to look after themselves. However in most of the developed and human-dominated world we have a major impact on the so-called wilderness, whether we like it or not. The first big problem is that the areas that we arbitrarily designate as wilderness or reserve are usually too small and too scattered to maintain stable populations of wildlife, and the bigger the animal, the bigger the problem. This can be offset to some degree by creating corridors along which terrestrial animals can move between suitable habitats (if they know the way). The next big problem, ecologically speaking, is that we are nearly always selective in terms of the species that we wish to preserve, be they orchids, bitterns or beavers. We must acknowledge therefore that most of the areas we call wilderness or reserve are areas that we have designated not to be used for food production but as habitats for species we favour, thereby adding to their non-use value to us as sources of beauty, recreation and the emotional satisfaction to be gained from the feeling that we are doing the right thing. What we cannot ignore is that, because we have created them, we owe them a duty of care.

There have been three approaches to conservation grazing.

- Intermittent short-term stocking with herbivores to 'tidy up' the vegetation.
- Permanent stocking with herbivores, with population numbers controlled by removal of animals for slaughter, or sale off the reserve.
- Permanent stocking with herbivores with population numbers subject only to 'natural' control.

In all cases it is necessary to select the animals most suited to sustain the quality of the vegetation, which means, in effect, the ones likely to do most good and least harm. Sheep are excellent in this regard, unless overstocked, because they concentrate on the grass and tend to leave the trees and shrubs alone. They are also the best distributors of fertiliser, wagging their tails to scatter their faecal pellets. Small hardy breeds of cattle (e.g. Galloway) that can subsist on highly fibrous grasses will never overgraze to the extent of ruining the pasture because of their eating behaviour. They scythe the grass with their tongues against their lower incisors so are unable to graze it to the ground. Sheep, on the other hand, under extreme grazing pressure will eat grass down to its roots. Cattle defecate indiscriminately but in wet lumps, from which grow untidy patches of rank uneaten grass. Moreover their feet can do great damage to the pasture ('poaching') in wet conditions. Horses especially ponies and other equines (e.g. zebras) equipped with both upper and lower incisors thrive very well on large, dry areas of grassland (savannah and steppes). In restricted areas they can graze grasses down to the roots.

The practice of intermittent short-term stocking with herbivores to 'tidy up' the vegetation is extremely effective when done well. On the dairy farms of south-west England, where I live, it is customary to bring in pregnant ewes from other farms to tidy up the mess left by the dairy cows when they are brought in for winter housing. In the mountains and moorlands, hardy cattle are brought in to preserve the environment for the hill walkers. Similarly in Switzerland, the presence of the cattle in the hills during the summer is vital to keep them open for the tourists.

The practice of permanently stocking nature reserves with herbivores, with population numbers controlled by removal of animals for slaughter, or sale off the reserve, can be highly effective and highly humane when done well. The primary purpose of the animals, maybe a combination of cattle and sheep, or cattle and ponies, is to maintain the quality of the habitat, and the cost of this can be met from subsidy and entry fees from those who come to enjoy the habitat.

However it is both economic and humane to control population numbers of animals, partly to generate income from the sale of meat and breeding stock, but also to maintain animal numbers at a level commensurate with the health and welfare of the herd.

The practice most likely to generate severe problems for the welfare of the animals and the habitat is the third option: permanent stocking with herbivores with population numbers subject only to 'natural' control. This has, in the past, created major welfare problems, most conspicuously (to my knowledge) at the 'Oostvaardersplassen' reserve on one of the reclaimed polders of the Isselmeer in Holland. This area of 56 km^2 was developed primarily as a reserve for waders and other birds. A policy decision was taken to 'manage' the habitat, grass and shrubs, with herbivores, hardy breeds of cattle, deer and ponies. The aim was to allow these species to establish an unmanaged equilibrium in terms of population numbers from 'natural wastage'. For the grazing animals, which, unlike the birds, could not fly away, and without the benefit of predators higher up the food chain in the form of wolves or omnivorous humans, population control was achieved through mass starvation, accompanied by massive destruction of habitat. As the grass was grazed to the ground, the cattle suffered most because they were unable to scythe the grass. The deer and ponies fared marginally better but, in so doing, killed the shrubs by stripping their bark. When I visited the area in March 2004, it resembled the Somme. Animals were being culled only when they were so weak as to be unable to stand. I understand that culling policy has become more intelligent since. This cautionary tale, once again, rams home the message that once we elect to interfere with nature we are obliged to assume the duty of care.

Sustainable intensification

It is a fact universally acknowledged that agriculture should be sustainable. Moreover most would agree that global sustainability should be defined both in terms of food security for all the people and the protection of the living environment. When we address the practicalities of how this might be achieved, we come across a wide range of views, which is perfectly natural since there is a wide range of options. At one end of the spectrum we have the low-input, low-output organic movement, who identify the conservation of land resources as a primary aim, but whose methods could not possibly provide all of the people with the food they would like to eat. At the other end, we have the 'sustainable intensification' movement who identify increased efficiency of resource use for mass production of food as a primary aim, while seeking to mitigate effects on the environment.

According to Jules Pretty (2008), sustainable production system should seek to pursue most or all of the following actions.

- Use crop varieties and livestock breeds with a high ratio of productivity to use of externally and internally derived inputs.
- Avoid unnecessary use of external inputs.

- Harness agro-ecological processes such as nutrient cycling, biological nitrogen fixation, natural predation and parasitism.
- Minimise use of technologies or practices that have adverse impacts on the environment and human health.
- Make productive use of human capital in the form of knowledge and capacity to adapt and innovate, and social capital to resolve common landscape-scale problems.
- Quantify and minimise the impacts of system management on externalities such as greenhouse gas emissions, availability of clean water, carbon sequestration, biodiversity, and dispersal of pests, pathogens and weeds.

These are all admirable aims but not necessarily compatible with intensification of agriculture. John Beddington has described the present crisis of food, water and energy insecurity as a 'perfect storm'. By his definition the provision of enough decent food for an ever expanding, ever more urbanised, ever more carnivorous population will require further expansion of industrialised, large-scale food production in a manner that does not seriously compromise our needs for the two other prime resources of water and energy. This is sound, indeed self-evident, so far as it goes, but it is still a limited (and selfish) aim since it defines security exclusively in terms of human needs. At present, most thinking on sustainable intensification is concentrated on further intensification of already highly intensive systems for mass production of food. This is realistic since if we are to provide enough good food for the global human population, then the majority of the food will have to come from highly intensive, highly efficient systems. It does not follow however that the majority of the land should be devoted to intensive cropping for food for humans and feed for animals. Indeed the aim should be to *reduce* the amount of land needed to support a defined quantity of meat, milk or eggs.

Debates about sustainable intensification are particularly susceptible to the fallacy of argument from limited premises (p. 7). One thing is certain: the apparent path to progress is dictated by where you stand at the outset. An international food company managing over one million laying hens will not approach the sustainable intensification of egg production in the same way as a community in an African village. While the principles of efficiency, complementarity, humanity and sustainability should be common to all systems, the correct application of these principles requires careful and sensitive analysis of the biological and socioeconomic circumstances of every individual scenario.

Industrialised animal agriculture

Application of the principles of sustainable intensification to industrialised agriculture is considered first because, in a strictly quantitative sense, it is by far the biggest issue. It has achieved its current highly productive state through the application of science and technology, and it is only realistic to assume that further progress will be achieved using the same tools. To date, most of the scientific argument has been directed towards improving yields and efficiency of food crop production. There has been relatively little constructive review of the role of the farm animals in sustainable intensification. For example, the 2009 report from the UK Royal Society deals exclusively with crop production, concentrating, as you would expect, on the role of new science, including genetic modification. The report acknowledges that in developed countries 70 per cent of food crops may be fed to animals, yet does not consider the extent to which overall efficiency may be improved by improving the utilization of this 70 per cent, measured in terms of efficiency, complementarity, humanity and sustainability.

When the terms of reference are strictly restricted to increasing yields and efficiency, especially through genetic manipulation, it makes sense to concentrate on crop production. As discussed in previous chapters, further improvements in the productivity of intensively reared farm animals achieved through improvements to feeding, breeding and environmental control are likely to be small. Molecular-based genetic engineering of farm animals to increase productivity has proved both highly unpopular and highly unsuccessful. Indeed, the trend in conventional breeding has moved towards the establishment of selection indices that are biased less towards simple productivity traits like growth rate and milk yield and more to traits related to robustness and lifetime performance. In the context of planet agriculture, the greatest potential for improved productivity in disease-free, intensively reared farm animals is to exploit and improve the nutritive value of feed from grasses and crop residues, thereby reducing the relative amount of arable land used to grow cereals and oilseeds for animal feed rather than human food.

Overall, for both plants and animals and in all environments, the most promising approach to sustainable intensification through improved yield and efficiency is to reduce losses from disease. In regard to most infectious diseases, there is still much that can and will be done by development of new vaccines and resistant breeds and strains of animals through the application of conventional or new genetics. As discussed in Chapter 7, there are promising signs that selection for genetic resistance to disease may not be restricted to single gene resistance to specific antigens, but encompass broad-spectrum polygenic resistance to complex disorders, such as respiratory disease in feedlot cattle, associated with multiple pathogens and environmental factors. How far this approach may succeed in reducing common but complex production diseases not strictly linked to infection, such as many of the conditions that cause lameness in poultry and dairy cattle, remains to be seen. However, so far as the animals are concerned, it would be no more acceptable to control infectious disease through genetic selection than through mass medication with antibiotics if, in so doing, it made it possible to sustain productivity in conditions of filth and squalor.

Animal agriculture in the third world

The capacity for sustainable intensification of animal production in villages and small farms in the third world is obviously far greater than in industrialised agriculture since one is starting from a much lower baseline. The UK Government Office for Science has examined the potential for sustainable intensification in African agriculture and proposed ways and means to bring it about. Their priorities for food production from animals may be summarised thus:

- Better disease management: e.g. breeding for trypanosome resistance in cattle, vaccination against Newcastle disease in poultry, improved parasite control.
- On-farm cultivation of new sources of fodder: e.g. perennial legumes, fodder shrubs.
- Cross-breeding to increase yields without compromising adaptability to the environment and disease resistance.

Clearly, subsistence farmers cannot be expected to do all these things without external assistance. This requires money, of course. It also requires education and training in necessary skills and care to ensure that all forms of support get directly to the people who are likely to put it to best use, usually the working women. Almost always this means giving support to the village communities, not the national government. There have been some conspicuous successes in this regard. Oxfam's Rakai chicken project in Uganda (Roothaert *et al*. 2011) established a poultry improvement programme based on local practices and cross-breeding indigenous chickens. Partnerships for implementation were strategically chosen. Scaling out was realized through participatory approaches, use of trainers from the communities, establishment of the Rakai Chicken Breeders Association, commercially and locally managed Integrated Feed Centres, market orientation and independent financial services, collaborating with and influencing local institutions. The average increased income as a result of improved poultry keeping was US$1050 per household per annum. This is an excellent extension of the Oxfam principle of 'teach a man to fish and you feed him for life'. In this case 'teach a woman to keep chickens and you feed a community for life'.

Sustainable intensification: downstream

The most common criticism of industrialised animal agriculture is that it wastes resources and pollutes the environment: wastes energy and organic matter, and pollutes the environment with (inter alia) nitrogen, phosphorus and bacteria pathogenic to humans. This, like most forms of blanket criticism, is seldom entirely rational. The extent to which different animal farming systems may act negatively or positively on the environment needs to be considered on a case-by-case basis, using a wide range of measures to create a life-cycle analysis (Tables 2.4, 2.5). I repeat: the poison is in the dose. On a traditional farm or a well-managed modern organic farm, the organic and mineral compounds in animal excreta are necessary to fertilise the land: self-contained organic farming cannot operate without animals. However, on a full life-cycle analysis, the production of meat, milk and eggs on a modern organic farm is always less efficient in terms of the utilisation of feed energy and usually less efficient in terms of overall energy use (feed plus fuel) than highly intensive industrialised livestock units, mainly due to lower output relative to the overall costs of maintaining the animals.

Advocates and apologists for industrialised farming argue that intensive livestock production can be environmentally friendly not only because feed conversion is more efficient but also on grounds of less methane emission from ruminants and overall a reduction in the amount of land allocated to food production. This argument is only valid when measured by the single criterion of efficiency. It neglects the values of humanity and stewardship. Extensive, well-managed grassland systems grazed by ruminants and equines are net carbon sinks (Table 2.5). Rearing animals in factory farms undoubtedly reduces the amount of land needed to grow monocultures of cereals and protein-rich oilseeds. However it ignores the extent to which farm animal production can contribute a major, but not the sole source of income within the sensitive, sustainable stewardship of the land for the benefit of all.

Anaerobic digestion

As a very crude first approximation, some 20–30 per cent of the matter and energy consumed by farm animals is passed out in their excreta. Industrial livestock farms are being increasingly encouraged to look on this as a resource, rather than a waste product.[1] The operation of a typical anaerobic digester is illustrated in Figure 8.2.

Manure or slurry from the animal unit is fed into the digester, to be fermented by anaerobic bacteria to produce combustible methane (CH_4) and CO_2 (as in the rumen, see p. 41). Conditions are optimal when the dry matter (DM) of the material is about 15 per cent and temperature 35°C. It may therefore be necessary to add heat and water, e.g. on beef cattle feedlots where the DM content of manure may exceed 50 per cent. Ideally this water should be 'brown' (contaminated) water from the unit that must not be allowed to enter watercourses. Fermentation for 20–30 days produces gas (CO_2 and CH_4), which can be stored. After removal of hydrogen sulphide this gas can be used to generate electricity for use on the premises or sale into the National Grid. After further removal of CO_2 it can be fed into national gas pipelines. About 50–60 per cent of solids are converted to biogas. The residue, at 5–15 per cent DM, contains all the original P and N, which can be spread as fertiliser in the form of slurry, or enter a further separation process to separate waste water from high DM manure. The process also removes over 90 per cent of potential pathogens from the excreta (Sharvelle and Loetscher 2010). Only about 10–15 per cent of the energy from cattle or pig excreta is captured as biogas, which doesn't sound very efficient. However there can be a lot of it: energy generated from a 1,000-head herd of dairy cows should be sufficient to heat 20 homes, or provide most of the power for the farm enterprise.

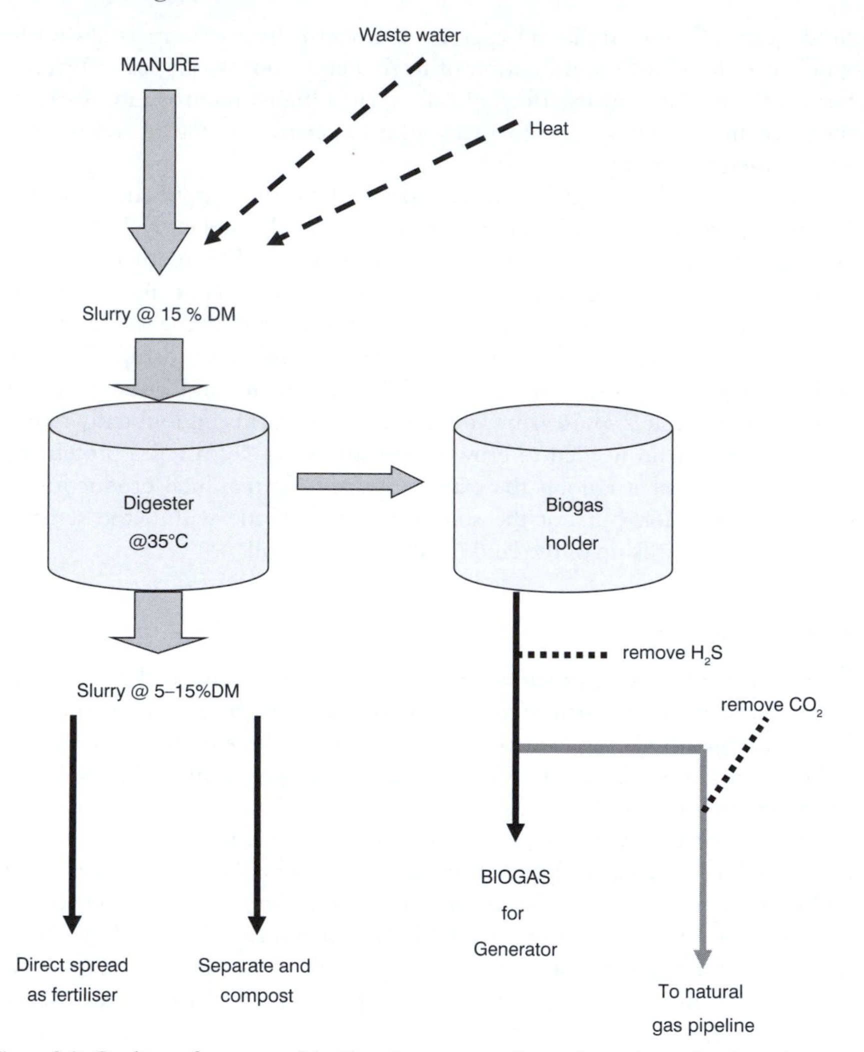

Figure 8.2 Outline of an anaerobic digestion process for an intensive animal production unit

Small-scale agriculture

While industrialised large-scale agriculture is now responsible for feeding most of the people, most of whom live in urban environments, most of the land and most of the people who work on the land are involved in very small-scale agriculture, with low inputs of machinery, fuel and capital investment. I avoid the term subsistence agriculture since I believe that all who work full-time or nearly full-time on the land deserve to be rewarded not only with food and clothing but also with sufficient income to give them the freedom to meet (at least) their basic needs for goods they cannot produce for themselves. However this is a book on husbandry, not third–world economics, so in this section I shall concentrate on the biological and behavioural interactions between small-scale farmers, their animals and the land.

At the subsistence level, the mutual dependence of humans and animals is total. It is simply not sensible to think of the animals as a luxury that they could do without or, at least, use more sparingly. The terms of reference should be the same as for any system of animal husbandry: efficiency, humanity and sustainability. There is practically no simple village system that could not be improved according to these criteria through a strategic and sympathetic combination of local knowledge and outside help both in the form of education and financial aid. Outside help can take many forms, some of which have been discussed already, such as control of infectious diseases and genetic improvement programmes appropriate to the local environment. Within the communities the greatest advances are likely to come through education, in particular education directed towards the better care of the animals, in every sense. A particularly fine example of this approach is provided by *Sharing the Load*, a guidebook to the practical care of working animals (van Dijk *et al.* 2011), sponsored by the Brooke Charity. This develops a progressive interactive approach to the improved understanding and communication of what is necessary to ensure the welfare of working horses, donkeys, buffalo, etc. and thereby to get a better return from a fitter, happier animal. It presents the evidence and outlines educational programmes and action plans at two levels: the keepers of the animals and the community facilitators. Welfare assessment and the practice of good husbandry are explained in simple English but in sufficient detail. The key information is also presented in cheerful pictograms intelligible to those who do not read English or cannot read at all.

This approach to education at the community level can meet the needs for improved efficiency and humanity. It does not address the problem of sustainability. Subsistence farmers are less to blame than anyone when it comes to squandering of resources and global warming. However they are prone to degrade the living environment in their immediate vicinity, to their own detriment in the longer term, usually because it is the only way they can survive in the short term. Much has been written about how to work simultaneously towards improved productivity and sustainability in third-world agriculture (Government Office for Science n.d.; Roothaert *et al.* 2011). At the biological level, there is general agreement as to what should be done. The politics and economics are not so straightforward.

It is also necessary to address issues and problems associated with small-scale agriculture in the developed and relatively affluent world. Here we can identify two main groups:

- traditional small-scale family farms resisting the general industrialisation of agriculture and urbanisation of society;
- new affluent hobby farmers, 'middle-class peasants', whose income can support a country lifestyle that involves part-time work on the land, and almost always keeping a variety of farm animals, since animal husbandry is usually viewed as the most fun part of the game.

The traditional small family farm that milks a few dairy cows and keeps a few sheep will usually qualify for high marks in terms of humanity and stewardship. In most cases they don't do much harm. However, they cannot compete with industrialised agriculture when measured in terms of efficiency of production of food as a commodity. Some of the more entrepreneurial and energetic have enhanced their own quality of life through schemes to produce and market added-value and niche products, where these are identified by measures such as local provenance, novelty (e.g. local cheeses), artisan production methods and high welfare (e.g. local grass-finished beef). This is exciting and the market would appear to be nowhere near saturation.

There is however another category of small farm that has been in the family for generations, where farmers have seen their role as simply to produce food and keep the land in good heart to pass on to the children. Fifty years ago this was easy. Before going to university I worked on a typical farm: 100 ha, a mixture of arable and pasture that supported 30 milking cows. The farmer employed four full-time workers, did no farm work himself except at hay time and harvest, put three children through fee-paying schools, hunted twice a week and maintained a regular supply of good whisky. The same farm today would need at least 100 cows to make ends meet. The farmer may work 14–16 hours/day with almost no days off. His children may see no future in farming on this scale and leave home to seek a different life. If suffering is defined as the inability to cope without difficulty with the challenges of life, then these farmers are suffering.

Relatively affluent hobby farmers can afford to give their animals individual care, including veterinary attention as necessary. However serious problems of animal welfare can arise on such farms, attributable to ignorance or neglect, or both. There is no excuse for neglect. However, some animals are harder to care for than others. Chicken, ducks and geese are relatively easy to keep, allowing for some natural wastage to predators such as foxes. Horses or ponies are almost obligatory for middle-class peasant families. These animals are likely to receive a great deal of human attention (at least while the children are at home). However, these are animals that evolved to live in herds and graze short-grass pastures for ten hours/day. They can suffer greatly when isolated in stables and fed on easy-to-handle pony nuts. Goats are excellent company but far too clever to be allowed free range. Sheep may appear superficially to be self-sufficient but are highly susceptible to external parasites, lameness and accidental injuries. They are seldom suitable for smallholders. The most recent edition of the UFAW manual *The Management and Welfare of Farm Animals* (2010), though intended primarily as a textbook for students of agriculture and veterinary science, can also serve as a practical introduction for novices seeking to embrace the delights of animal husbandry.

Stewardship of the countryside

This section comes with a warning: it is both parochial and polemic. We, the British, love our countryside, with some justification because much of it is beautiful indeed. However, to revert to my main theme, we don't value it enough. Once we step outside our gardens, we entrust its care to others and do little more than grumble when that care fails to come up to our expectations. As for wildlife, a hardy few seek it in nature reserves, the majority are content to watch them go through their repertoire of killing and copulation on television. While we relax in the comfort of our homes, the British countryside is fighting for its life against the destructive and conflicting demands of all those who are using it for their own limited ends, motivated by the desire for profit and the need to survive. Large arable farmers profit from strip-mining large areas of land to produce crops for food, feed or biofuels. This may be necessary but, to our eyes, it is not beautiful, and to all the other plants and animals excluded by these monocultures, it can be a disaster. Meantime the traditional mixed farmers, the stewards of the 'green and pleasant' remnants of our countryside, cannot sustain a living from what they see to be their job, namely feeding the people. As they give up, or withdraw quietly to shoot themselves, developers are attracted to the scent of a 'green-field site' and move in to feast on the spoils.

We may draw some comfort from the increasing attention being given to nature reserves, conservation areas and sites of special scientific interest. These are admirable, not just for our sakes; more so for the sake of the plant and animal life that they sustain. However in the UK, with the exception of a few sites, mostly in Scotland, these conservation areas tend to be small, isolated, and seriously under-funded. Moreover they are not true wilderness, but sites managed by conservationists with a 'vision', to recapture an image of the country as it was 100 years ago, or in the mesolithic period (9,000 years ago), to preserve the fritillary butterfly or to reintroduce the beaver. This all may be very nice but it is not natural and, on this scale, it may not even be sustainable.

Most of the British countryside is farmed land but much of it, especially the most beautiful parts, cannot support itself simply from the production of food and other elements of 'use value'. Stewardship of the living environment demands proper respect for, and should give proper value to, all life on all the land, plants and animals, wild and domestic, farmers who own the countryside, townies who *need* the countryside. Wherever possible the distinction between farmland and wilderness should fade. The aim should be a sustainable balance of nature between domestic and wild flora and fauna, whose overall value is defined by their utility, their beauty and their contribution to the health of the living planet.

Let me give two very different examples of how farmed and wild animals can coexist provided that the stewards of the land are properly rewarded for their stewardship. In areas such as the English Lakes and the Scottish Highlands, the habitat that we now value for its wild beauty has largely been sculpted by grazing sheep. Of course, values change and the present appearance of the Highlands would look like a desert to the crofter, thrown off his land 250 years ago to make way for the great white sheep that could make more money for the estate owners than the crofters, largely from the sale of wool. Today the price of wool barely covers the cost of shearing and the sheep can no longer pay their way. Farmers brainwashed to believe that their only path to salvation is through increased efficiency (in producing commodities) sought solutions in increased production (more sheep), leading to accelerated degradation, loss of biodiversity and very little increase in income. Conservation site managers in the National Parks now seek the cooperation of farmers to reduce grazing pressure and, where this is happening, we are seeing recovery of biodiversity, shrubs, tall herbs, alpine flora and the birds and insects that thrive in these habitats. The problem, however, is that society (i.e. us) is not yet prepared to pay enough to do this properly.

My home in Somerset (south-west England) is in traditional pasture-based dairy cow country. Quantitative audit of agricultural systems (Chapter 2) shows this to be both a highly productive form of agriculture and highly efficient by virtue of its use of complementary feeds. As such it should be sustainable simply from the sale of food. However, as discussed in earlier chapters, pressure to increase income from the sale of milk has driven farmers to select for high-yielding Holsteins that are unsuitable for pasture-based systems and have to be kept in barns and fed high concentrate rations. When income is dominated by the sale of dairy produce, these large zero-grazing units outcompete the more natural, humane and sustainable pastoral systems. The market ignores (not entirely) the invaluable contribution of the cows to the sensitive and sustainable stewardship of the British countryside. Where I live, many of the cows are at pasture for eight to nine months of the year. They are moved through small fields with high hedges rich with nesting birds. The lanes between the fields explode with wild flowers. When the cows come in for milking, they bring in the muck but the muck attracts the insects and the insects attract the swallows and martins that nest in the eaves and the swifts that soar, and sleep, in the sky.

If society at large is to enjoy the beauty and amenity of the countryside then the sheep farmer in the hills of Scotland who earns very little from sale of animal products deserves considerable reward for stewardship of his land for the benefit of the community at large (humans, wild animals and the whole living environment). The dairy farmer operating a traditional grass-based system earns considerably more than the extensive sheep farmer from the sale of milk, particularly if he is an active partner in downstream exercises in adding value through the production of specialist cheeses, ice cream and yoghurt. However if he elects to farm in wildlife-friendly fashion, with some loss of potential income from milk sales, then he too deserves some (lesser) reward.

The Common Agriculture Policy from the European Commission was established in 1957 with the aim of increasing agricultural productivity, partly for food security reasons, but also to ensure that the EU had a viable agricultural sector and that consumers had a stable supply of affordable food. By 2003 it had become apparent to all that this was not the best way to spend a great deal of money since it had created vast amounts of surplus food and totally failed to address matters of humanity and stewardship. There has however been progressive revision of the CAP that has (in my opinion) gone some way but not far enough to redirect support away from production and towards stewardship of the environment and the support of rural communities. At present support is based on two pillars.

- Pillar 1 is designed to provide direct support to individual farmers. Originally it was linked to provide market support (i.e. supplement income from sale of produce). In recent years this has progressively shifted towards a single payment, decoupled from production but subject to the condition that the land continues to be farmed subject to 'cross-compliance', measured in terms of the extent to which farmers meet standards of environmental management and (to a lesser extent) animal welfare.
- Pillar 2 is intended for Rural Development Programmes. These are seen as an important mechanism for delivering biodiversity and other environmental objectives including greenhouse gas mitigation in the agricultural sector and rural areas.

In 2008 the total European Agricultural Guarantee and Guidance Fund (EAGGF) was close to €50 billion, over 50 per cent of total EU expenditure. Direct, uncoupled payments to farmers accounted for 67 per cent, payments coupled to production 15 per cent. Only 18 per cent has been allocated to Pillar 2, Rural Development (DEFRA 2006).

It is neither possible nor relevant to examine here all the policy decisions that govern the allocation of funds to individual farms and member states within the European Union. It is all too easy to criticize a policy that allocates billions of euros to farmers in uncoupled payments and broadly in proportion to the size of their farms. The policy of cross-compliance is sound in principle but weak in practice since it demands little more from farmers than concordance with regulations. The probability of failure to meet standards for cross-compliance is vanishingly small. Nevertheless, the present CAP policy, however clumsy, is helping to keep farmers on the land without creating food mountains. Moreover it permits some flexible interpretation by the member nations and stays reasonably clear of micromanagement.

The EC continues to keep the CAP under review and recently set out a series of options for future reform (EC 2012). Option 3 is remarkably sympathetic to the central theme of my argument.

- Pillar One, direct payments: phase out direct payments in their current form. Provide instead limited payment for environmental public goods.
- Pillar One, market support: abolish all market measures with the potential exception of disturbance clauses that could be activated in times of severe crisis.
- Pillar Two, rural development measures: to be focused mainly on climate change and environmental aspects.

These broad aims are, in my opinion, very sound. However there is much work to be done. We need clear definitions of what should be included within the category of environmental public goods. We need clear direction as to what measures to mitigate climate change and enhance environmental aspects should receive support. We then need a careful economic assessment of the lasting value of each of these environmental goods, which will determine the distribution of CAP funds between the various actions designed to meet these needs. In my view the aim should be to direct all the subsidy available for support of agriculture (leaving provision for emergency support in times of crisis) away from production and towards the stewardship of the living environment, on the sound basis that society should be prepared to pay at the time the full cost of the food we eat but that the long-term needs of the living environment are better served through strategic planning and funded by taxation. Even if the current CAP budget was halved this would still be a lot of money. At present in the UK there are Environmental Stewardship schemes that offer small amounts of money to encourage farmers to make their farms more wildlife friendly by extending hedgerows, etc. However many farmers consider the rewards too small to justify the effort (and loss of production). Moreover they focus only on the prettier features of planet husbandry, like wildlife, and neglect the big issues such as carbon sequestration and water management. Farmers and landowners need big rewards if they are to do big things for the planet.

Overall, the financial support of the living environment needs to come from a variety of sources: sale of food and farm goods, including added-value products such as specialist cheeses. Government subsidy (taxpayers' money) is necessary when the rewards are not immediately apparent to the people, such as habitat protection, carbon sequestration and water management. However in areas where society at large can perceive (or be educated to perceive) the non-use value of specific goods and services, like amenity and gold standards of production and animal welfare (e.g. organic, freedom foods), then it is fair that individuals should pay. Compassion comes at a price.

Wildlife and disease in farm animals

One of the greatest constraints to the vision of an Eden wherein farmed and wild animals coexist in conditions of comfort and security is the problem of transmission of infectious diseases. This problem acquires even greater importance when the disease is zoonotic, i.e. transferable also to humans. The most obvious way to avoid this risk is to keep them apart. In traditional, extensive, nomadic and tropical communities this may be unrealistic, although there have been great efforts to keep domestic cattle free from trypanosomiasis by keeping them apart from wild ruminants and the tsetse fly carriers. The situation is different in areas of industrialised, intensive animal farming such as Europe and North America where it is practically possible to isolate farm animals from wildlife reservoirs of infection by rearing them in confinement, even though this may be less than satisfactory in terms of animal welfare and sustainable land use.

Table 8.1 lists in alphabetical order the most important diseases transmissible between wild and farmed animals within temperate zones where farm animals are likely to be reared in confinement. Influenza describes a set of important zoonotic diseases involving various subtypes affecting many species, most

Table 8.1 Important diseases transmitted between wild and intensively farmed animals (for a comprehensive global picture see Pastorete *et al.* 1998)

Pathogen	*Farmed species*	*Wild species*	*Human risk*
Avian influenza	Poultry	Wild birds	Yes
Foot and mouth virus	Cattle, sheep, pigs	Wild boar, buffalo	No
Leptospirosis	Cattle, sheep, pigs, horses	Rodents, deer, hedgehogs	Yes
Newcastle disease	Poultry	Wild birds, semi-domestic pigeons	No
Salmonellosis	Poultry, cattle, sheep, pigs	Wild birds, many wild mammals	Yes
Swine fever, classical	Pigs	Wild boar	No
Swine fever, African	Pigs	Wild boar, warthog	No
Tuberculosis, bovine	Cattle	Badgers, deer, many wild mammals, wild birds?	Yes

critically poultry, pigs and humans. The most important strain of avian flu in terms of both bird and human health is H5N1, which affects both domestic poultry. For most of the world the disease is epidemic rather than endemic. However there is a risk of transmission to outdoor poultry flocks (and thereby to humans) from infected wild birds migrating from areas where the disease is present. As with all serious epidemic diseases of farm animals, disease-free status is the ideal state and sometimes, not always, this is best preserved by a policy of eradication by mass slaughter. Vaccines are available in areas where the risk is high. In Europe, the risk from wild birds is low and the times of risk are predictable. At periods of high risk, it would be prudent to keep free-range hens shut up but there is absolutely no case for prohibiting free-range hens on grounds of health and safety. The same principles apply to Newcastle disease, highly infectious and highly pathogenic for birds but harmless to humans and effectively controlled by vaccination. Salmonellosis does present a problem for free-range poultry since it can be transmitted from the excreta of wild birds that take feed from outdoor feeders set out for the hens. The obvious solution to this is to place the feeders inside the house. However this reduces the incentive for birds to go out at all. One option is to site the feeders in a covered veranda area between house and field. Having got as far as the veranda, birds are more inclined to venture further.

Foot and mouth is a highly infectious disease (FMD) of cattle, sheep and pigs that is rigorously controlled by eradication where possible or repeated vaccination, where not. There have been outbreaks of FMD in Europe arising from contact with wild boars but the risks are low. Leptospirosis is a serious zoonotic disease that affects a wide range of farmed and wild species. Control can be achieved through a policy of vaccination supported by attention to rodent control on farm. It does not call for any radical changes in husbandry. Salmonellosis is also a serious disease with a wide host range, including humans, and involving a very large number of different strains. The management of salmonellosis on farm and as a matter of public health is a complex issue and outside the remit of this chapter. However it is fair to conclude that wildlife are minor contributors to risk.

The badger dilemma

One of the greatest areas of conflict in regard to the problem of disease transmission between wild and farmed animals revolves around the issue of bovine tuberculosis (bTB), especially in the UK. In the first decades of the 20th century this was the most important zoonotic disease in the UK. In the 1920s 40 per cent of dairy herds were infected with bTB and thousands of people, mostly children, contracted tuberculosis as a result of drinking infected milk. This route of infection has been effectively controlled by pasteurisation. There is still a small risk to humans drinking unpasteurised milk or working in direct contact with infected cattle, but bTB is now essentially a cow problem. In the interests of improving the health and international trading value of the British cattle population it became from 1950 the policy of the State Veterinary Service to eradicate bTB from the national dairy herd through a compulsory policy of routine diagnostic 'tuberculin' testing, slaughter of infected animals and restriction of movement of cattle on infected premises. In the early 1980s it appeared that the eradication process was close to complete success (infection present in less than 0.4 per cent of herds). However in the last 20 years the situation has been deteriorating at an alarming and accelerating rate, particularly in areas such as south-west England where there has also been a very large increase in the population of badgers, many of whom are infected with bTB. In 2008 infection was present in 6.9 per cent of herds, 36,000 cows were slaughtered and the cost to the UK taxpayer had risen from £25 million in 1989 to over £100 million. Once again, a full explanation of the epidemiology and control of bTB in cattle is outwith the scope of this chapter (see Ward *et al.* 2006).

It is however an inescapable if unpalatable fact that wildlife, especially badgers, constitute a major hazard. This presents a massive problem of both ethics and law since it involves a conflict between the welfare of the badger and the welfare of the cow. In the UK it is an offence in law 'wilfully to kill a badger, to interfere with a badger sett, or to disturb a badger when occupying a sett' (Protection of Badgers Act 1992). The State Veterinary Service has a statutory duty to eradicate bTB in cattle and this would undoubtedly be helped by a draconian policy of killing all badgers in areas where the disease was endemic. The statutory duty of society to refrain from killing a badger or disturbing a badger sett wilfully ignores a major problem in public health. The two instruments of policy cannot coexist and something has to give.

Government policy in the UK, after years of dithering, appears to have settled on a strategy based on selective slaughter in areas of high infection supported by vaccination in surrounding areas. This policy, supported by controls on cattle movement and strategies to keep cows and badgers apart, should help to reduce the incidence of bTB in both badgers and cattle.

The badger dillemma presents an extreme example of conflict of interests between ethics, economics and emotions as experienced by different individuals. To return to one of my main themes: animal welfare is important but not all-important and has to be accommodated within an ethical and political framework that respects all parties. The extrinsic value of the cow is greater to the farmer and has a far greater use value for most of society than the badger. So far as the animals are concerned, the untimely slaughter of 36,000 visually healthy cows per annum far exceeds the cost of killing badgers, many of which are severely ill. However the badger carries a greater extrinsic value than the cow in relation to the emotional needs of some of society. In this context the badger is akin to the tiger. Most of us will never see a live one in the wild and would not wish to bump into one on a dark night. Neither species is of any use to us but we would feel the world to be a poorer place without them. All I can say to such people is that there is no intention to exterminate the badger. Our responsibility to the badger is no more nor less than our responsibility to any other species of sentient wild animals that coexists on farmland. We should, with humanity and stewardship, seek to sustain population numbers that are compatible with the overall fitness of the species, while seeking to minimise the suffering of individuals. In some cases the active humane killing of individuals is likely to do more to alleviate suffering than a 'hands-off' policy that substitutes the Malthusian controls of starvation and disease.

Futures for farm animals: conclusions

This chapter concludes the case for the continuing constructive contribution of farm animals to future agriculture, considered in terms of biological efficiency, food quality and human health, animal health and welfare, ethics and economics and finally the overarching concept of planet husbandry. The approach, thus far, has been based on science, in particular, quantitative biology, since almost nothing in the whole argument has been identified in absolute terms as 'good' or 'bad'. The sustenance, or the poison, is in the dose, which means that any form of advocacy is likely to be facile unless it acknowledges the evidence and gives due regard to the numbers. The final chapter will consider how we, the people, individually and collectively, can help to work towards these aims. Here I set out the key issues, matters of fact and matters of conjecture that need to be taken into account when working towards this end.

- A stable ecosystem involves a balanced interaction between plants and animals. This applies equally to farmed land and to wilderness.
- Much animal farming has seriously compromised the stability of ecosystems: intensive farming through pollution, extensive farming through degradation of land and soils.

- Most of those who can eat too much food of animal origin in terms of individual health, collective food security and environmental sustainability.
- The mass production of arable crops for feeding to farm animals adds to the inequities of food availability for humans, and loss of forests and other ecosystems necessary for climate stability.
- Man's exploitation of the farm animals has consistently failed to give proper regard to their sentience and capacity for suffering. Large-scale, industrialised animal farming is not inherently different in this regard with respect to animal populations but is systematically unable to treat individual animals with due respect.

However

- When animals are fed on complementary rather than competitive feeds, they can improve the overall efficiency of production of highly nutritious food for humans. This can apply at all levels from village chicken production to intensive dairy farming.
- Areas of grassland, or mixed grass and woodland, can make sustainable contributions to immediate and long-term value (e.g. food and C sequestration respectively) but each contribution needs to receive its proper reward.
- The quality of animal husbandry, measured in terms of animal health, welfare and land use, including the protection of wildlife habitats, can be improved without significantly increasing the cost of food. This can be encouraged by the direction of subsidy towards these elements of 'non-food' value.

But

- Action for change is likely to be most effective when carried out at the scale of the individual farm or village. However, overall policy must be directed at planet husbandry and the pursuit of this aim *must* include a progressive reduction in the relative amount of resource allocated to the feeding of farm animals.

Note

1 In traditional mixed farms it always was.

9 Animal husbandry and society

Substance and shadows, carrots and sticks

The philosophers have only interpreted the world in various ways. The point is to change it.

(Karl Marx)

The story so far has been concerned with the husbandry of farm animals and the land involved in the farming of animals. Husbandry is soundly based in science and economics but enriches these things with three special human qualities: duty, care and conservation. The three principles that have been used to define good husbandry are efficiency, humanity and stewardship. The impact of animal husbandry has been measured in terms of the productivity, health and welfare of the farm animals and the living environment. The quality of husbandry is directly determined by the actions of those who have a direct duty of care for the animals and the land, i.e. the farmers. The main aim of the story so far has been to help those who are directly involved or plan to become directly involved, or advise those who are directly involved to act as well as possible.

Respect for the duty of care to the farm animals and the farmed environment must extend to all who derive value from the animals and the land in the form of goods and services. This means all of us. The few who reject the assumption of dominion over the animals, such as vegans and Jainists, are expressing this principle of respect in the most profound form. This final chapter therefore considers the responsibility of all society to all that live off the land – farmers, animals and the whole living environment – not just (to paraphrase Karl Marx, above) in terms of what should be done but how it can be done.

Moral principles

I start from the premise that it is acceptable to rear and kill farm animals to provide us with food and other goods and services. Since my behaviour is consistent with the overwhelmingly omnivorous majority of human kind I shall not attempt to marshal a moral argument in favour of animal husbandry; it is a fact of life. Previous chapters have examined areas where animal husbandry can make positive and negative contributions to planet husbandry. One thing however is certain: it can't and won't go away.

Application of bottom–up ethics to this large fact of life requires us, the moral agents, to assume responsibility for the animals and the land, our moral patients. In the case of the domestic animals that we use for own ends (this applies to pets as much as to the food animals), our responsibility is to seek a fair and humane compromise between what we take, in the form of food or companionship, and what we give, in terms of good husbandry, i.e. competent, humane care. According to this one-sided concept of justice, we can only assume the right to control their life and death if we accept the responsibility to give them a life worth living.

In order to get the best from our sentient animals, measured in terms of productivity and efficiency, and do our best for them, measured in terms of our humanity and their welfare, we need an empathetic understanding of their physiology and behaviour; how they work, how they feel and how they behave. It follows that those who seek to get the best out of individuals within society, measured in terms of action (rather than just sympathy) for the welfare of farm animals and the living environment, must base this on a similarly empathetic understanding that we humans are also sentient beings whose motivation to action is based at least as much on how we feel as on what we think – or, in many cases, attempts to rationalise how we feel on the basis of thoughts selected to favour our (sentient) preconceptions.

Individual human attitudes to the management of the food animals and farmed land extend across a very broad spectrum. One end of the spectrum assumes complete freedom of action to do anything (or neglect to do anything) to an animal that is permissible in law. This has been the moral position of the majority for most of the history of agriculture although, of course, human attitudes to what is and is not permissible have evolved over time and this tends (eventually) to be incorporated into changes in the law. At the other extreme of the spectrum we discover individuals with a passionate concern to protect animals from all forms of human interference, sometimes matched by an equally passionate desire to condemn all humans who don't think the same way. In the middle ground between these two extremes we can recognise two distinct groups. The first very large group consists of those who are not directly involved in the business of food production from animals and give little or no thought to its provenance, either because they simply don't care, or because it upsets them to think about it – so they don't. The second group, small but growing, consists of those who are not directly involved in the business of providing goods and services from farm animals but derive benefit from these goods and services, are troubled by what we are doing to our animals and to our planet and are prepared to make a positive contribution to doing things better. It should be obvious by now that you are a major target for my argument in favour of animal husbandry regained and a critical force for putting it into effect.

The several and sometimes complex arguments presented in this book, based on audit and action to improve efficiency, humanity and stewardship, are intended to achieve consistency between what we feel in our hearts *ought* to be done and what the evidence suggests *should* be done. It would however be both arrogant and unnecessary to suggest that everybody who wishes to contribute to the resurrection of animal husbandry should attempt to acquire a working knowledge of all the issues. Scientists and those politicians who profess a faith in science tend to dismiss emotional expressions of abhorrence at some of the things we do to animals as unreliable – too unreliable to be a basis for legislation, although powerful enough, in a democracy, to stir our elected politicians into a semblance of action to suggest that something will be done, if not yet.

The most common criticism of the emotional approach to our treatment of animals is that it is rooted in anthropomorphism. We ask ourselves 'how would this chicken/mouse/monkey feel if it were me?' I believe that rational scientists should think twice before condemning anthropomorphism out of hand. When an uninformed person expresses horror when first shown a battery unit for laying hens or an intensive fattening unit for pigs, then it is fair to question the system. Indeed Frances Wemesfelder has revealed a great deal of agreement between novices and experts when assessing the welfare state of pigs just by looking at them (Wemelsfelder *et al*. 2001). This may be because we can empathise with pigs. It would, I suggest, be more difficult for a novice to assess the welfare state of a lobster. Animal behaviour scientists do, in fact, practise a form of 'reverse anthropomorphism'. They turn the question around and ask 'how would I feel if I were this pig/mouse/monkey?' To give a specific example of such a question the ethologist might ask 'If I were a pig would it make me feel bad (e.g. frustrated) if I had no access to a foraging substrate? If I could relieve this frustration through access to a foraging substrate, what would I prefer?' The next step is to convert this thought experiment into a set of scientific experiments designed to pose these questions to the pig in a way that it can understand. The scientific term for this reverse anthropomorphism is called motivation analysis. The responsibility of the scientific method compels us to accept the outcome of the experiment, properly designed on the basis of a null hypothesis. To put it more simply, we must trust the choices and the actions of the pig, not our own prejudice.

In many cases, the outcome of carefully controlled scientific trials will do no more than confirm the first impressions of a novice: pigs really are happier when they can forage outdoors or in deep straw than when confined in a barren environment on concrete slats (Wemelsfelder *et al.* 2001). In other cases the results of scientific trials can be counterintuitive. Batteries of 'furnished' cages for laying hens may still appear profoundly unappealing to the uninformed observer but there is good scientific evidence that they can meet most of the hen's important behavioural needs (Appleby *et al.* 2002). They achieve no less satisfactory a compromise between the conflicting demands of the Five Freedoms than many free-range units.

If we are to progress towards better, kinder animal husbandry then it is necessary to seek a compromise reasonably acceptable to all parties within the ethical matrix: producers, consumers and farm animals. The public must be happy to buy it, the farmers happy to do it and the animals happy to accept it. Only thus can we change the world.

Interlude: a pig's eye view of human behaviour

If we are to do right by the pig, then we should try to empathise with how it must feel to be a pig. To this end I offer an extreme exercise in reverse anthropomorphism: to imagine how a pig might imagine it must feel to be a human. Wittgenstein suggested that if a tiger could talk we couldn't understand it, presumably because of profound differences in the nature of our consciousness. However I see no harm in trying. The pig is a good choice for this exercise because, as Churchill said, 'dogs look up to you, cats look down on you, a pig treats you as an equal'.

> I, the pig, start from the premise that humans have as much right as me to be considered a sentient animal. Sentient animals have feelings that matter. I, the pig, assume that because humans, like us, are sentient you are powerfully motivated, like us, to behave in ways designed to make you feel good and avoid feeling bad. From what I observe, this can involve a wide range of behaviours, some of which I can understand, seeking food, comfort, sex. Others that I just don't understand include running marathons, adopting stray dogs, sitting indoors all day discussing the meaning of words like deontology when the sun is shining and a wallow is at hand.
>
> Being a pig, I am a libertarian but one with respect for the golden rule of 'do as you would be done by' (which, if I thought about it, is presumably what you mean by deontology). I therefore acknowledge that you humans should have the freedom to do what you like so long as your actions do not frighten the horses or compromise the freedoms of others. Because I am a well-read pig I describe these as negative freedoms, after Isaiah Berlin (1967). Positive freedoms are only permissible up to a point. In the neat phrase of George Monbiot, 'your freedom to swing your fist stops at the point of my nose'. So far as I, the pig, am concerned you have an obligation to respect four of the five freedoms you have suggested for us, namely the negative freedoms from hunger and thirst, thermal and physical discomfort, pain, injury and disease, fear and stress. The fifth, positive freedom to exhibit normal behaviour, is, like your positive freedoms, negotiable. I share with you the grudging acceptance that I should not be free to copulate whenever I like with whoever is at hand. Having reached this point I, the pig, conclude that this is about as far as I can go. I humbly concede that, while you humans appear to share with me the property of sentience, feelings that matter, most human behaviour is so alien to my understanding of the world as to be fundamentally uninteresting or constitute a threat (I too read Wittgenstein). Thus, on balance, I decide that my contact with humans should be kept to the bare minimum consistent with my personal interest.
>
> Ah, you will say, not all animals think and feel like pigs. Sentience and cognitive ability will be governed by species, evolution, education and environment. We can agree on this. I think we can also agree that wild animals will sensibly conclude that they should stay out of your way and you out of theirs. We observe that domesticated farm animals you have

made dependent on you appear to interact with you in a positive way. However this may simply be because they anticipate you are bringing them something they need. Remember the neurobiological experiments of Keith Kendrick (1998) with sheep. The image of a human triggers a signal of aversion. The image of a human carrying a sack of corn triggers a signal of attraction. You humans should recognise that, while we may respond positively to you when you are being useful, you have no right to expect us to love you. So long as I am in a decent habitat and with the company of my own kind, contact with humans is not a behavioural need.

I would ask you to extend this concept to relationships with your favourite pets, dogs, cats and horses. Remember they are not the same: dogs have owners, cats have staff. Cats enjoy your company but on their terms. When did your cat last stroke you? Dogs may appear to love you but that is because most of them are not fully aware they are dogs and that's a problem. Dogs and horses exhibit a broader repertoire of behavioural disorders than any species of farmed animal raised in the company of others of its own species. The biggest abuse to the welfare of dogs and horses is that they are denied freedom of natural behaviour achieved by social contact with their own kind. This is in direct conflict with the UK Codes of Practice for the Welfare of Farm Animals. If dogs and horses were not classed as pets, this denial of natural behaviour could be considered an abuse.

To conclude: I concede that you have made yourselves necessary to us, but you remain a pretty inexplicable species. The bit that I can understand is that you, like us, are driven by your sentience and motivated largely by self-interest. This may cause you to feel love for certain animals, individuals and species. However this does not give you the right to expect us to love you. It follows that the best interactions between humans and other animals are likely to be working relationships based not on emotion but on mutual respect for each other as individuals: cowboy and horse, shepherd and dog, peasant and house cow. I am not remotely concerned as to how you feel, it is what you do that matters.

You may find all this hard to take, and indeed, profoundly disagree with me; but then you are not a pig. I rest my case.

Pathways to right action

Progress in regard to the welfare of farm animals and the living environment depends on right action by individuals and society in general. Within this broad context right action is, I repeat, that which seeks a fair compromise between the conflicting needs of the different parties. This does not require a descent into moral relativism since it is entirely consistent with the absolute moral principle of justice.

All individuals share the responsibility for right action. However our individual motivation to right action is likely to be strongest when it is consistent with our individual needs and perceptions of value. Most individuals are likely to recognise the use value of such things as wholesome, tasty food and the non-use value of a beautiful countryside. Many individuals are also likely to be motivated by the altruistic wish to do good by others: humans, animals and the living environment. However, without meaning to be cynical, I suggest that for most good people the strength of motivation towards actions that bring immediate personal benefit is likely to be stronger than towards actions designed for the general and long-term good. The instruments of a just society, (i.e. good government and the law) have the responsibility to promote action for the general good, especially when it brings no immediate reward to the individual, or conflicts with individual self-interest.

The rules of engagement for right action within the context of animal husbandry are defined by:

1 A clear definition of animal welfare ('fit and happy') and a systematic approach to its evaluation (the 'Five Freedoms').
2 A structured and comprehensive understanding of the interactions between farm animals and the environment measured by way of transactions in matter and energy (audits of agriculture including life-cycle analysis).
3 A sound ethical framework that affords proper respect for the value of farm animals and the living environment within the context of our duties as citizens to the welfare of human society.
4 Realistic, practical, step-by-step strategies for improving the quality of husbandry, measured by its impact on farm animal welfare and the farmed environment, within the context of other, equally valid aspirations of society.
5 An honest policy of education that can convert human desire for better production standards into human demand for better products.
6 Quality control in respect to systems of food production from animals based on robust protocols for assessing animal welfare and the provisions that constitute good husbandry linked to quality assurance schemes that promote increased individual demand for these added-value products.
7 Legislation by proscription and incentive (stick and carrot) to promote right action, especially in regard to the environment, when it is unlikely to be achieved through individual action.
8 'Politics by other means': action by individuals and groups within society to encourage and promote husbandry standards that are above that required by existing laws and regulations and encouraged to aim even higher.

The first four rules of engagement have been covered in previous chapters. Here we need to consider, first, how individuals within society may be directed honestly towards a better understanding of the farmers' need to improve production systems and our need to reward them for their effort. We then need to create a basis for trust that farmers are doing the right thing, through effective quality assurance. Finally we need to examine what governments, non-governmental organisations and highly motivated individuals can do, should do and should not do to improve animal husbandry through legislation, education and persuasion.

Awareness and understanding, images and reality, education and propaganda

Any attempt to persuade humans to change their behaviour towards farm animals and the environment (or anything else!) will fail unless it is based on a rational understanding of the complex factors that influence human behaviour, and a sympathetic understanding of human response to these factors, i.e. maximum appearance of the carrot, minimum application of the stick. One does not need to go into the complexities of human psychology in order to recognise and acknowledge the strength of the influences that are at work.

Throughout most of the history of civilization, indeed until the second agricultural revolution in the 20th century, most people had some practical experience of farming and farm animals. Unless they were directly involved in farming, their understanding was likely to have been limited and, as for concern, most probably did not much care. Nevertheless, their images of farming systems, however superficial, would have been based on the evidence of their own eyes. Today most of the people in the developed world, including many of those with strongly expressed opinions on animal welfare and the impact of farming on the environment, have little or no experience or education in respect to these things; their views are based on images rather than reality.

Images, by definition, are no more than snapshots of reality. Moreover most are created to convey a preconceived message. Some preconceptions are likely to be more balanced than others but, as I stated at the outset, all arguments are made from limited premises. The main sources for communication of images are:

- communications directed at a general audience: television, radio, newspapers, online (e.g. Wikipedia);
- publications, written and online, designed to promote special interests: farmers, retailers, environmental groups, animal welfare societies;
- uncontrolled media: Facebook, Twitter;
- word of mouth.

Television, radio and newspapers are, in most parts of the world, subject to some degree of control and surveillance of standards. Nevertheless, communications within these media are prepared to meet a set word count and a tight deadline by journalists with many other things to think about, and an agenda that seeks to maximise public appeal by provoking shock and horror or, at least, accentuating the negative. I have great respect for a few journalists but I would not be so naïve as to expect them to present a comprehensive and balanced picture. (Comprehensive and balanced pictures can be less than gripping, especially when propounded by scientists.)

Communications prepared to promote special interests (e.g. farmers' groups, animal welfare charities) are biased by definition. This does not necessarily condemn them. In my opinion the great majority of publications from groups representing farmers, retailers, respectable welfare charities and other nongovernment organisations are honourable and seek to be fair. Their clear aim is to emphasise their own side of the argument. On the whole however they tend to be prepared with care, not least because each conclusion that goes on the record becomes a hostage to fortune in the form of contradiction from the other side. New, unregulated social media such as Facebook and Twitter may produce the occasional gem but have the potential to disseminate any old rubbish. I would not include them in my list for further reading.

The most powerful mechanism for further broadcasting of images generated within the regulated media is word of mouth. A major food retailer explained to me their company policy in response to the impact of a television programme drawing attention to a new food scare (e.g. salmonellosis in poultry). Following a chicken scare of this sort the supermarket will reduce supply to meet a short-term decline in demand, the extent of which is calculated from previous experience. I was advised that during the school vacation the number of people influenced by the message that appeared on television is five times the number of people who watched the programme. In term time, the multiplier is 20 in consequence of four times as much interaction between mothers when they meet at the supermarket having dropped their children off at school. Moreover the duration of the drop-off in sales is likely to be a matter of a few days or weeks. This expression of human behaviour was seen even in the case of 'mad cow disease' (nvBSE). Sales of beef recovered most of the lost ground within weeks, which is hardly logical for a disease where the incubation period may, on occasion, be well in excess of 20 years.

Education has to be the best form of vaccination against contagion from images that present, at best, an incomplete picture and, at worst, can be dangerously wrong. The classical origin of the word is *e ducare*, which means 'leading out': guidance towards a state of understanding through unbiased presentation of the evidence, structured analysis of the evidence on the basis of science and humanity, and policy for action based on justice to all parties worthy of respect. All education should be a lifetime experience and the more the better, but any education, in my belief, is a good thing. It is often said that 'a little knowledge is a dangerous thing'. This can only be true of knowledge (or images) acquired without understanding. Proper education should convey both knowledge and understanding but it should also instil awareness of how much we do not know and how little we truly understand. This provides a shield against both wrong belief and wrong action. Knowledge and education concerning the husbandry and welfare of animals kept in the service of man can shield us from irrational belief in propaganda from any source, be it industrialists, politicians, welfare lobbyists, or oddballs. Knowledge and education should also restrain our own impulses to wild assertions.

Because this book deals with such a wide range of topics I have only been able to present a very small segment of the knowledge required to acquire a reasonable degree of understanding in each. What I have tried to do is to introduce and outline the big issues of science, economics and morality within the three areas of efficiency, humanity and stewardship. For those who want or need to explore these issues in more depth I have provided just enough suggestions for further reading to get you started. Thereafter you are on your own.

While it is neither necessary nor possible to educate everybody in the nature of everything relating to animal husbandry, each individual, or administrator of other individuals, needs enough knowledge and understanding to ensure that the actions for which he or she is responsible are, indeed, responsible actions. There are three levels at which this education should be addressed:

- general awareness;
- education for the public good;
- education for the professional.

I have already identified increased awareness of the nature of animal sentience as the single most effective step towards increasing the sum of human compassion for other life forms. This can be instilled early in life. Young children can be brutal because the principle of 'do as you would be done by' does not appear to be innate. However children are motivated to love their pets. From this excellent starting point they can, with some effort, be motivated to care for their pets and, from this, motivated towards a compassion for all sentient animals. Adults, even in the most brutalising environment of the abattoir, can be moved to treat animals better simply by bringing the matter to their attention. I was very impressed to read the following notice written over the point of entry of animals to a Scottish abattoir: 'Quality control starts here. Animals should be treated with care and compassion at all times.' If such a sign had been present twenty years ago it would almost certainly have ended with the word 'care'. Simple awareness of the need for compassion addresses the moral virtue of respect for life.

Education for the public good is that which is necessary to guide the actions of those who benefit from animals but do not actually work with them. The first step in this process is to seek the closest possible match between how people feel about problems of animal welfare and the environment and the hard evidence on which our actions should be based. This will help to ensure that action in response to public demand will, in fact, be right action. This addresses the moral virtue of beneficence. The next step is to convey knowledge and understanding in respect to the benefits we derive from the animals and how these benefits to us may be achieved at least cost to them. This addresses the moral virtue of justice. As always, such education cannot be enforced simply through the didactic presentation of 'facts' to be learned and accepted. Didactic teaching can only be justified as a method for preparing the mind to interpret personal experience. If children are to acquire a sympathetic understanding of farm animal welfare and the impact of farming on the environment, this needs to be developed through a process of directed self-education to convey knowledge and understanding of what goes on and why. One of the best ways to achieve this is to make as many farms as possible open to the public according to a structured programme constrained only on grounds of health, safety and proper respect for the farmer's need to earn a living without constant interruption. Farm visits are particularly valuable for schoolchildren, ideally just prior to puberty when their minds are as yet uncluttered by information, propaganda and emotional turmoil. These visits should be formally linked to teaching and testing within the curriculum in matters of food production, food safety, environmental protection and animal welfare. They should be much more than a jolly day out in the country. This can only be achieved when education authorities provide the funds and the time within the curriculum to meet this essential element of education in citizenship.

The third category is education of the professionals, those who make their living by working in direct contact with the animals and the land. Here the goal is professional competence and this involves a lot of work. While the main conclusions of this book are intended to contribute to understanding at all three levels, the details are aimed primarily at the trainee and active professionals.

Educated eating

For those of us who inhabit the relatively affluent, relatively well-fed world, one of the major contributors to right action is to eat less food of animal origin. This can win the triple crown: sustained personal health, fairer sharing of resources with the hungry poor, and better health and welfare for the living planet. Some of the evidence that commands this conclusion has been presented in previous chapters. The problem at issue here is how to bring about change: to educate the people towards healthier eating habits. The problem is confounded by the fact that one cannot simply regard 'the people' as a unity to be engineered towards a single concept of right action. Happily we are all individuals with individual beliefs, wants and needs. Many individuals simply don't care what they eat so long as they enjoy it. Such individuals need to be encouraged and educated to care a little more, if only in the selfish interests of their own health and the fitness of the offspring that will sustain their genetic line. Others identify food and health as a major source of anxiety and are therefore exquisitely susceptible to false images, food fears and the aggressive marketing of specialist diets and food supplements usually in the absence of good evidence and often in flagrant contradiction to the evidence. These people need to be encouraged to understand a little more and worry a good deal less.

One of the major roadblocks on the path to improved education in human nutrition is that, in most countries, there is no professional body similar to that which exists for other disciplines related to health, like human and veterinary medicine, that sets standards of competence both for practising nutritionists and the teaching of nutrition. In consequence, anyone can claim to be a nutritionist and establish a career from the sale of dubious diets and superfluous supplements to the anxious well. The UK has recently established the Association for Nutrition that defines the necessary competencies and monitors the necessary standards for the education and continuing professional development of registered nutritionists. At present there is no legal requirement to restrict the title of nutritionist to those who can meet these criteria, which means anyone can still claim to be one. However the register does make it possible to distinguish between the qualified consultant and the quack.

Quality assurance and quality control

Those directly involved in the business of animal husbandry, whether farmers, researchers, advisers or administrators of the law and welfare have a primary duty to promote good farming standards. All parties in the food chain beyond the farm gate – producers, retailers and consumers – should also seek to adopt and promote best practice. This can only occur if all parties receive a just reward for right action. This reward may take the form of increased sales or increased profit margin, although this is not the only measure of reward. What is certain is that no one can be driven so hard to do the right thing that it puts them out of business.

Active workers in the business of producing food from animals, farmers and farm workers, veterinarians, hauliers and abattoir workers, butchers and dairy workers, all require formal education, including directed self-education in the practicalities of the job to convey the knowledge and understanding appropriate to their special functions. In the context of good husbandry, understanding implies both the ability to interpret knowledge gained from didactic teaching in the light of their own practical experience, and a compassionate respect for the animals for which they are responsible. This takes time. I repeat: it is very easy to care *about* animals. Caring *for* them requires skill and experience.

The great majority of people in the modern developed world have no direct contact with farm animals so cannot be expected to acquire the specialist knowledge necessary to understand all the rights and wrongs of animal husbandry. It is right that they should have concern for the welfare of farm animals and the environment and better still if they have a rational awareness of the key issues, but nobody can expect to be fully informed about everything. To put it another way, all of us are ignorant, we are just ignorant about different things. Thus we have no option but to take most things on trust. It falls to those who are directly concerned with animal husbandry and the production of food from animals at all stages from farm to fork to ensure that this trust is well founded.

Farm assurance schemes have been developed for most livestock sectors in the UK and Europe (e.g. DEFRA, Assured Dairy Farms, RSPCA Freedom Foods, Soil Association). Different quality assurance (QA) schemes place different emphasis on food safety, animal welfare and the environment. The primary purpose of a QA assessment is to ensure compliance with the standards of the scheme. As a minimum any QA scheme must include all legislation that is relevant to the stated objectives of the scheme. However the degree to which the public may be prepared to pay more to reward a 'green' or 'high-welfare' scheme will depend on the extent to which it is perceived to improve on minimal standards. A scheme whose standards are designed to admit any farmer who doesn't actually break the law is not likely to appeal to the discerning consumer.

There have been two basic approaches to the implementation and promotion of 'high-welfare' standards. The RSPCA 'Freedom Foods' approach defines standards for husbandry and welfare that are considerably higher than the legal minimum, independently monitors compliance with these standards and promotes them on the basis of trust in the organisation. The alternative approach proposed for Europe by Welfare Quality® (2009) is to rank farms according to the score attained during welfare assessment as *unclassified, basal, good* and *excellent.* A similar ranking approach, the '5-Step Animal Welfare Standards', has been developed in North America by the Global Animal Partnership (2008). The attraction of these welfare-labelling schemes is that they allow for (and encourage) continuous improvement. They are fundamentally sound in so far as the standards of animal welfare required for compliance are based on sound principles of science and good husbandry. Their success will depend on their impact on consumer behaviour and first results are promising. Sales of 'Freedom Food' free-range eggs in the UK rose from under 10 per cent to over 50 per cent over the period 1996–2010. Sustained progress will require a build-up of consumer trust in the QA claims of the scheme, based on transparency: the knowledge that the evidence relating to standards of animal welfare and environmental stewardship is consistent with the claims of the scheme and available for inspection.

The virtuous bicycle

Delivery of a QA scheme that simultaneously adds value to food from animals on the basis of quality control on farm and increases consumer demand for these added-value products requires coordinated action on farm and beyond the farm gate. My proposed approach to this is based on two virtuous cycles of effective action: the 'Producer Cycle' and 'Retailer Cycle', which together make up a 'Virtuous Bicycle' (Figure 9.1; Webster 2009). This was originally designed strictly within the context of animal welfare but can readily be expanded to incorporate environmental management.

The four steps in the producer (or 'farm') cycle are:

1. Self-assessment of husbandry standards by the farmer: evaluation of resources, records, health plans, etc.
2. Independent audit of husbandry, health, welfare and environmental management: based on evaluation of the self-assessment and observation of welfare outcomes.
3. Action plan prioritised to give attention to principal hazards and critical control points.
4. Reassessment and review of action plan.

The initial self-assessment by the farmer (based on a structured but not too rigid questionnaire) has several merits: it can reveal both farmer knowledge and attitude, it can provide much more information than will normally be acquired

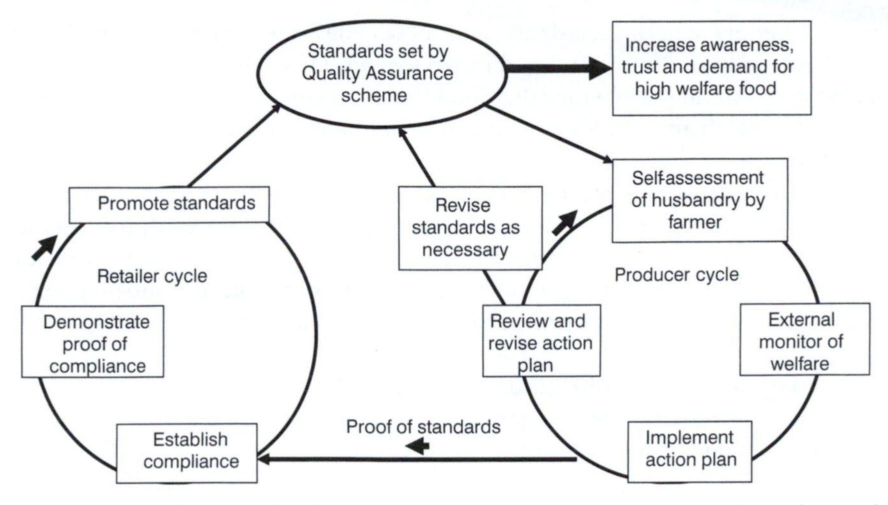

Figure 9.1 The Virtuous Bicycle, a delivery system for improved husbandry and animal welfare (Webster 2009)

during an annual quality control (QC) inspection lasting less than half a day. Moreover, it saves time. The first QC inspection by the independent, trained auditor will pronounce whether or not the farm is in compliance with the standards of the QA scheme or how it ranks within the scheme. The farmer should have the right of appeal against this initial assessment. The independent QA assessor may identify and prioritise needs for a dynamic action plan for health and welfare although s/he should not be directly involved in drawing up the plan. The action plan should be formulated according to HACCP (hazard analysis and critical control point) principles (see p. 183) that identify the most important hazards and call for action at the most critical control points. The practical benefit of this approach is that it focuses the attention of the farmer on the things that matter most. In our experience these plans are more likely to be effective when the ownership is held by the farmer than by an external consultant such as a veterinary surgeon, for the simple reason that the farmer has the most to gain or lose.

The components of the retailer (or 'fork') cycle are:

1 Setting quality standards for husbandry and welfare: either absolute or ranked.
2 Demonstration of compliance with standards.
3 Promotion of products that demonstrably meet quality standards.
4 Review of standards and audit procedures in the light of experience.
5 Rewards for producers in compliance with quality standards.

Baseline quality standards for all farmers within UK or the European Community are defined by law and reinforced by welfare codes (e.g. DEFRA, www.defra.gov.uk/food-farm/animals/welfare). It is right that all farmers should comply with this legislation and these codes and it is necessary that they are enforced by audit from trained inspectors who have the backing of the law. However, it should be obvious that food that can be marketed only on the basis that it is compliant with legislation can never be classified as any higher than *acceptable*.

The main attractions of this approach to on-farm assessment are that it should involve much less repetitive 'box ticking' than most current QA protocols. It generates a dynamic action plan to improve husbandry that focuses on major issues, and calls for evidence of effective action at critical control points. However farmers are fully entitled to ask the following questions:

- 'Where are the rewards?' (money, praise, pride)
- 'Will it create real improvements in welfare?'
- 'Will you ever concede that I am good enough?'

All these are valid concerns. No farmer will be keen to enter a voluntary QA scheme that loses money or makes him less competitive. However, it would be insulting to suggest that the only motivation for farmers to enter a high-ranking scheme is to increase financial return. Pride in work has always been at the core of good animal husbandry but this pride deserves reward in the form of overt recognition that the farmer is doing a good job. Reward therefore depends on a fair price for high-welfare food, sustained by a lasting contract.

One of the major concerns about current QA schemes is that there is, as yet, little evidence to suggest that they are delivering what they claim, i.e. significantly higher standards than on farms not participating in the scheme. One perfectly acceptable reason for this is that standards can be satisfactory on many non-participating farms. A second reason, that gives more cause for concern, is that current QA schemes do little more than audit farms on an annual basis, confirm compliance (or not) and go away. Both auditor and farmer can then forget about it until next year. The virtuous bicycle differs in that it calls for and monitors the effectiveness of action at a realistically limited number of control points. However, this too is easier said than done. It is not sufficient to define and demand these standards; we need to explore the incentives and constraints to getting things done. These are illustrated in Figure 9.2, which explores the motivation to take effective action to deal with threats, e.g. threats to individual health and welfare, or threats to the health and welfare of animals and the land on an individual farm. Motivation is defined by the perceived magnitude of the problem, or threat, and perceived effectiveness of possible actions to remove or reduce the threat (Webster 2012).

To illustrate the elements of farmer motivation as outlined in Figure 9.2, consider the major welfare problem of lameness in dairy cows. The farmer has first to acknowledge the magnitude of the threat. This can present a major obstacle as many dairy farmers seriously underestimate both the severity and prevalence of lameness in their herds. Cows that are hopping lame are likely to be recognised, though not necessarily treated. However the cow with a

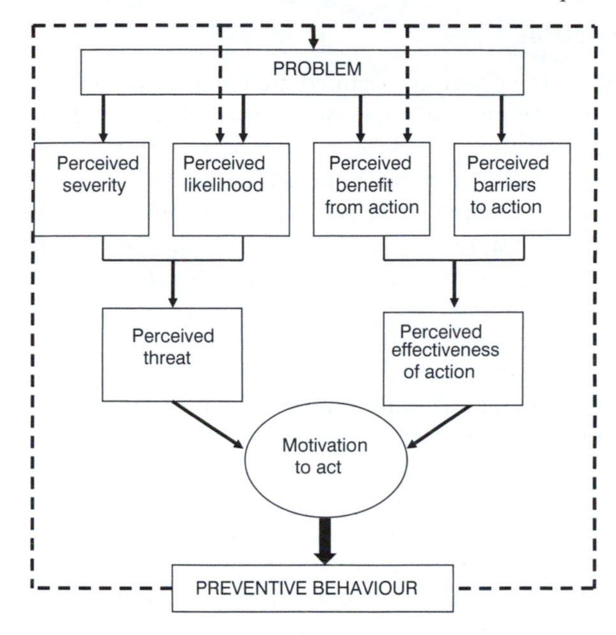

Figure 9.2 Farmer motivation: incentives and constraints (Webster 2012)

slow, hesitant gait has, on many dairy farms, come to be seen as normal. The perceived effectiveness of action depends on the balance between the perceived benefits of action and the perceived constraints to action. All dairy farmers, I am sure, would like to see less lameness in their dairy cows. However many are likely to be discouraged by the perceived barriers to action (e.g. the cost of radical reconstruction of the cow accommodation) relative to the perceived benefit measured in terms of increased income. There is however good new evidence to indicate that the most important hazards for dairy cow lameness relate not to housing or nutrition but to standards of foot care (Bell *et al.* 2009). Foot care takes time, and time is a scarce commodity for dairy farmers. However there are no major capital costs so the perceived barriers to action are small. The effectiveness of the action can be measured in economic terms: increased milk sales, reduced veterinary costs, increased productive life for the cows. There can also be a quiet satisfaction in no longer having to walk behind a succession of limping, suffering stragglers en route to the milking parlour.

This is just one example of the factors that determine motivation to preventive behaviour as illustrated in Figure 9.2. The principle is identical in concept to the motivation to promote personal health through a change in eating habits, e.g. reduce the risk of colonic cancer by reducing consumption of red meat. In both examples the evidence on which to base the effectiveness of action is sound. I can however think of many examples where the evidence is not sound, e.g. the benefits to humans of organic food or vitamin supplements, the benefits to farmers of selecting animals for increased productivity. Where sound evidence is lacking we should be reluctant to offer advice.

Collective action

Farm husbandry standards are defined in the first instance by legislation, reinforced by regulations and codes of practice. Chapter 5 described in some detail the fundamentals of the law as it relates to the farming of animals and the necessity for the law. To recapitulate in brief: it sets absolute standards to which all must conform, it commands respect for non-commercial concerns that are outside the market and it conveys a sense of trust in the principles of the law without the necessity for everyone to examine all regulations line-by-line.

Prescriptive law

This can do no more than set minimum standards as defined by the legislation of the moment. It does not create incentives to develop standards that exceed the requirements of legislation. This, as suggested earlier, is more likely to be advanced by a combination of drivers: the market, as defined by the wishes of the people, and government incentives, e.g. the allocation of subsidies. The good news is that, in a just society, the law does evolve to reflect new evidence and the wishes of the people. The bad news is that these may not be the same. Let us consider some of the approaches to working for improved welfare through political action. The area that is most amenable to progress through politics by decree involves those industrialised systems of animal production (battery hens, broiler chickens) where welfare problems are intrinsic to the system and largely independent of the quality of stockmanship within the system. Honest politicians who respect the ethical matrix will seek to achieve a just compromise between the needs of the animals and the needs of the people. In practice this is likely to involve legislation for modest improvements in husbandry standards that generate modest increases in the minimum cost of food for all the people. Examples of such legislation are the ban on the pregnancy stall for sows and imposition of improved environmental standards for caged laying hens.

Modest increases in legislation for environmental standards that carry only a modest increase in cost to the consumer can do more for farm animal welfare (considered in the utilitarian sense) than more ambitious attempts to prescribe gold standards. While only a relatively small proportion of the community may be prepared to pay 50 per cent more money for niche food products from a small number of animals marketed as super green and super compassionate, legislation for improved standards for all animals that increased the price of food by only 5 per cent would be (a) unavoidable, and (b) largely accepted as fair. Most of us will concede that we need the help of the state to make us into better people, but not if it costs too much.

Legislation by incentive

This can create a climate of opportunity for improvements to the husbandry of farm animals and the environment that go beyond the minimum standards defined by proscriptive law. In Chapter 8 I discussed how the strategy for the Common Agricultural Policy (CAP) of the European Union has evolved from one primarily linked to production (and food security) towards one directed more towards creating incentives for elements of value uncoupled from production such as the welfare needs of the animals and the quality of the living environment (EC 2012). While I am personally enthusiastic about the thinking behind the future planning of the CAP, it is still not possible to give it a pass mark in its present form, when the largest amount of money is doled out to farmers simply for owning the land and roughly in proportion to the amount of land that they own.

The amount of money allocated to the CAP should, in my opinion, remain substantial, though not perhaps as high as at present. *None* of the money should be used to support production; individuals who benefit directly from the food they buy should be prepared to pay the full price, especially when they have great freedom of choice, determined according to circumstances by cost, convenience, appearance, taste (and many other etceteras). *All* or nearly all the money should, in my opinion, be directed towards the stewardship of the living environment. At present the allocation of funds for environmental improvements are relatively small but extremely limited in scope, e.g. creating habitats for wildlife. What we need is a far more comprehensive policy that gives due reward to farmers who contribute to environmental and climate stability, e.g. carbon capture and water conservation through well-managed agro-forestry schemes, and penalises both those who pollute and those who degrade.

There are many who argue that a significant proportion of the CAP should be directed towards improving standards of farm animal welfare. To the consternation of some I have to say that I can give only limited support to this appeal. It is right that subsidy from the CAP should be linked to standards of animal welfare that are in compliance with regulations and codes of practice. However I do not consider that international policies for legislation by incentive are likely to have a significant impact on the development of farms ranked as *good* or *excellent* when assessed in terms of animal welfare. This is more likely to be achieved through the power of the people, or 'politics by other means'.

Politics by other means

To recapitulate: legislation by prescription has the merit that it is enforceable. Legislation by incentive and paid for out of taxation can encourage actions that are necessary for us all, such as climate stability and the protection of the environment. Nevertheless, there are clear limits to what can be achieved through the conventional political process. I repeat: it is possible to legislate for improved minimum standards for caged hens or commercial broiler production because the production systems and the animals used within those production systems have become extremely standardised. However when we come to consider more traditional family farming systems, whether in the developed or developing world, legislation is not really appropriate in relation to anything beyond the most basic standards of animal welfare because different individual farms have different individual problems and these require specific solutions. Here the best approach to improvement at the farm level is through a QA system as described above, operating within the free market to standards defined (e.g.) by a supermarket chain or an independent body such as the RSPCA or the Soil Association.

The other problem with politics by decree is what David Fraser calls 'Demosclerosis'. Anyone who has sat for any time on committees drafting new legislation or codes of practice will have suffered from the slow erosion of good intentions through expedience. 'Must' becomes 'should'. 'Should' then becomes 'should, unless'. It is at times like this that action through politics by other means becomes particularly attractive. I am not here referring to banner and megaphone diplomacy. What I am referring to is action aimed directly at those who have the most power to promote or enforce higher standards of animal welfare. These are, of course, not the farmers but the major retailers who themselves demand standards of quality assurance from the producers who supply their goods and are themselves absolutely dependent on their own quality assurance measures, and image, if they are to maintain market share in an environment where the consumer has freedom of choice.

McDonald's, the burger barons, made a company decision to instigate improvements in animal welfare, particularly at the point of slaughter, that go far beyond that required by legislation. It may be that McDonald's had become morally aware of the need for beneficence. They may also have been commercially aware that a large sector of their market is made up of young people who have a greater concern for the welfare of farm animals than they do for their own health and may, when McDonald's are getting a bad press, choose to buy their burgers elsewhere. In either event, it is clear that McDonald's were pressed into action by the force of public opinion. When a company that claims to sell 67 million burgers per day acts to improve its welfare standards, life, or at least death, is made a little better for a very large number of animals.

In Britain, the public demand for free-range eggs has radically altered the buying habits of the supermarkets and this has fed through to the producers. The interval between the publication of the Brambell Report (1965) and the phasing out in the EU of the conventional battery cage for laying hens (2012) will be 47 years! The growth of free-range egg production in the UK, from 5 per cent to over 50 per cent in ten years, in response to the force of public opinion acting directly upon the retailers has simply driven straight through the slow march of the legislators.

Farms, where animals form an integral part of biologically sustainable systems for working the land, are more likely to prosper through 'politics by other means' than through legislation. The aim here is for farmers to work with the retailers to supply premium food products to consumers seeking added value, whether defined by high-welfare, 'organic' sustainability, local produce, or any combination of these three. As discussed already, for this approach to succeed, the customer must be able to trust the claims for added value made by the producers and retailers. It is unrealistic to expect that every discerning customer will make it their business to discover what goes on within every farm, or even read all the small print on the QA leaflets. However, that trust should be based in the knowledge that we can, if we wish, get access to the truth. This could involve direct access, a personal visit to production units or, more realistically, trust in the competence and honesty of an independent inspector acting on our behalf.

Politics by other means is undoubtedly a powerful force for change. Inevitably however it will be driven by the most vocal and charismatic, who are not necessarily the best to advise on the best way forward. The pressure for the ban on the battery cage came not from facts carefully accumulated by animal welfare scientists but from the image of the battered hen in the battery cage put forward by charities such as Chickens' Lib and Compassion in World Farming. In this case the scientists were able to provide the evidence that led, eventually, to a change in the legislation. In matters of politics by other means it is the duty of scientists to regard this righteous emotion with a sympathetic but cool eye: to help right thought to proceed to right action. As always, this will be achieved through good interactive communication designed to ensure that the heart does not get too far ahead of the head.

Justice for all

To conclude: the goal of my entire argument has been justice for all involved, directly or indirectly, in the business of animal husbandry. My motivation has been the British sense of fair play. The concept of justice extends not only to all (human) moral agents, producers and consumers, but to the moral patients for which we share responsibility, namely the farm animals and all that makes up the living environment. The absolute moral principle of justice for all within such a broad canvas commands the need for compromise: no party has the right to expect too much. Fortunately for the moral argument, it is consistent with our most basic and selfish needs for individual and collective survival. Most of those who can afford it can reduce the risks of disease and premature death by eating less food of animal origin. Those who cannot currently afford it deserve a fairer share of the meat, milk and eggs. For them nicer food is also healthier food (up to a point). The living planet *must* receive a more sympathetic and, above all, a more educated and intelligent approach to husbandry if successive generations are to enjoy a reasonable quality of life. I am not so apocalyptic as to forecast the extinction of the human race but life for us could become extremely unpleasant, not least because an environment in which almost every living thing has been subsumed to the feeding of the human race would be bleak in the extreme. I have seen the future, in the hinterland around Beijing, and it hurts.

The principles that govern our paths towards the goal of justice for all are efficiency, humanity and stewardship. Within each category one can identify specific paths to right action, e.g. (and respectively) complementarity in the selection of feeds for animals, environmental enrichment and the control of production diseases for intensively farmed animals, carbon sequestration in pastoral and agro-forestry systems. There will always be the need to pursue new scientific knowledge and understanding, and we will continue to be surprised by advances we never imagined. However, much of what we need to know, we know now. The obstacles arise when we try to put this knowledge into effect. If we exclude natural disasters, most of these obstacles can be attributed to deficiencies in human behaviour leading to failures in justice through individual selfishness, political weakness, short-termism or downright corruption. Despite all this I retain my faith in the belief that there are quite enough humans prepared to do the right thing by others so long as they are not overstressed by the demands for personal survival, and receive a fair deal in return. You are my target audience. I hope I have been able to help.

References

Alderman, G., and Cottrill, B. R. (1993) *Energy and Protein Requirements of Ruminants: An Advisory Manual Prepared by the AFRC Technical Committee on Responses to Nutrients,* Oxford: CAB Direct.

Appleby, M. C., Walker, A. W., *et al*. (2010) Development of furnished cages for laying hens. *British Poultry Science,* 43: 489–500.

Arbel, R., Bigun, Y., *et al*. (2001) The effect of extended calving intervals in high lactating cows on milk production and profitability. *Journal of Dairy Science,* 84: 600–8.

Beauchamp, T. L., and Childress, J. F. (1994) *Principles of Biomedical Ethics,* Oxford: Oxford University Press.

Bell, N. J., Bell, M. J., Knowles, T. G., Whay, H. R., Main, D. J., and Webster, A. J. F. (2009) The development, implementation and testing of a lameness-control programme based on HACCP principles and designed for heifers on dairy farms. *Veterinary Journal,* 180: 178–88.

Berlin, I. (1967) *Concepts of Liberty.* In A. Quinton (ed.), *Political Philosophy* (pp. 141–52), Oxford: Oxford University Press.

Blaxter, K. L. (1989) *Energy Metabolism in Animals and Man,* Cambridge: Cambridge University Press.

Bludell, J. E., Lawton, C. L., Cotton, J. R., and MacDiarmid, J. I. (1996) Control of human appetite: implications for the intake of dietary fat. *Annual Review of Nutrition,* 16: 285–319.

Botreau, R., Veissier, I., Butterworth, A., Bracke, M. B. M., and Keeling, L. (2007) Definition of criteria for overall assessment of animal welfare. *Animal Welfare,* 16: 225–8.

Bourn, D., and Prescott, J. (2002) A comparison of the nutritional value, sensory qualities, and food safety of organically and conventionally produced foods. *Critical Reviews in Food Science and Nutrition,* 42: 1–34.

Brambell, F. W. R. (1965) *Report of Technical Committee to Enquire into the Welfare of Animals Kept under Intensive Husbandry Systems,* Cmnd. 2836, London: HMSO.

Broom, D., and Johnson, K. G. (1993) *Stress and Animal Welfare,* London: Chapman & Hall.

Brown, L. R. (1995) *Who will Feed China? Wake up Call for a Small Planet,* New York: W.W. Norton & Co. and Worldwatch Institute.

Bywater, R., *et al.* (2004) A European study of antimicrobial susceptibility among zoonotic and commensal bacteria isolated from food animals. *Journal of Antimicrobial Chemotherapy,* 54: 744–54.

Casas, E., Kuehn, L. A., McDaneld, T. G., Smith, T. P. L., and Keele, J. W. (2011) Genomic regions associated with incidence of disease in cattle using DNA pooling and a high density single nucleotide polymorphism array. *Journal of Animal Science,* 89 (e-suppl. 1): 158 (abstract 18).

Cook, J. T., McNiven, M. A., Richardson, G. F., and Sutterlin, A. M. (1995) Growth rate, body composition and feed digestibility/conversion of growth-enhanced transgenic Atlantic salmon (Salmo salar). *Aquaculture,* 188: 15–32.

Corbière, F., Cassard, H., Foucras, G., Meyer, G., and Schelcher, F. (2008) Hepatic abscesses: pathogenesis, treatment and prevention in cattle. *Le Nouveau Praticien Veterinaire Elevages et Santé*, 9: 36–41.

Corry, J. E. L., and Atabay, H. I. (2001) Poultry as a source of campylobacter and related organisms. *Journal of Applied Microbiology,* 90: 96S–114S.

Danbury, T. C., Weeks, C. A., Chambers, J. P., Waterman-Pearson, A. E., and Kestin, S. C. (1999) Self-selection of the analgesic drug carprofen by lame broiler chickens. *Veterinary Record,* 146: 307–11.

Dawkins, M. S. (1990) From an animal's point of view: motivation, fitness and animal welfare. *Behavioural and Brain Sciences,* 13: 1–61.

Dekkers, J. M. (2004) Commercial application of marker- and gene-assisted selection in livestock: Strategies and lessons. *Journal of Animal Science,* 82: E313–38.

Department for the Environment, Food and Rural Affairs (DEFRA) (2006) *A Vision for the Common Agriculture Policy,* London: HMSO, www.defra.gov.uk.

Department for the Environment, Food and Rural Affairs (2011) *Food Statistics Pocket Book:* <www.defra.gov.uk>.

De Soto, H. (2000) *The Mystery of Capital: Why Capitalism Triumphs in the West and Fails Everywhere Else,* New York: Basic Books.

Dohoo, I. R., DesCôteaux, L., Leslie, K., *et al.* (2003) A meta-analysis review of the effects of recombinant bovine somatotropin: 2. Effects on animal health, reproductive performance, and culling. *Canadian Journal of Veterinary Research,* 67(4): 252–64.

D'Silva, J., and Webster, J. (2010) *The Meat Crisis: Developing More Sustainable Production and Consumption,* London: Earthscan.

EFSA (European Food Safety Authority) (2012) Guidance on food and feed RA from GM animals and GM animal health and welfare, www.efsa.europa.eu.

European Commission (2012) *The Common Agricultural Policy after 2013:* http://ec.europa.eu/agriculture/cap-post-2013/index_en.htm.

European Food Safety Authority (EFSA) (2010) The Community Summary Report on antimicrobial resistance in zoonotic and indicator bacteria from animals and food in the European Union in 2008. *EFSA Journal,* doi:10.2903.

Fitzsimmons, R. C., and Newcombe, M. (1991) The effects of ahemeral light-dark cycles early in the laying cycle on egg production in White Leghorn hens. *Poultry Science,* 70: 20–5.

Food and Agriculture Organization (1998) *Village Chicken Production System in Rural Africa: Household Food Security and Gender Issues,* FAO Animal Production and Health Papers, 142, Rome: FAO.

Food and Agriculture Organisation (2006) *Livestock's Long Shadow: Environmental Issues and Options,* Rome: FAO.

Fox, D. G., Sniffen, C. J., O'Connor, J. D., Russell, J. B., and Van Soest, P. J. (1992) A net carbohydrate and protein system for evaluating cattle diets: III. Cattle requirements and diet adequacy. *Journal of Animal Science,* 70: 3578–96.

Frederiksen, B. S. (1997) Legislation in response to the Nitrate Directive aspects for some EU countries. In F. Brouwer and W. Kleinhanss (eds), *The Implementation of the Nitrate Policies in Europe: Processes of Change in Environmental Policy and Agriculture* (pp. 43–60), Kiel: Wissenschaftverlag Vauk.

Gardner, C. D., *et al.* (2007) Comparison of the Atkins, Zone, Ornish and LEARN diets for change in weight and related risk factors among overweight premenopausal women. *Journal of the American Medical Association,* 297: 970–7.

Gardner, I. A. (1997) Testing to fulfill HACCP (hazard analysis critical control points) requirements: principles and examples. *Journal of Dairy Science,* 80: 3453–7.

Gardner, I. A., Parsons, A. J., Xue, H., and Newman, J. A. (1997) High sugar grasses: harnessing the benefits of new cultivars through growth management. *Proceedings of NZ Grassland Association,* 71: 167–75.

Gibson, J. P., and Bishop, S. C. (2005) Use of molecular markers to enhance resistance to livestock disease: a global approach. *Scientific and Technical Review – International Office of Epizootics*, 24: 343–57.

Global Animal Partnership (2008) The 5-step programme, www.globalpartnership.org.

Government Office for Science (UK) (2011) *Sustainable Intensification in African Agriculture: Analysis of Cases and Common Lessons,* www.bis.gsi.gov.uk/foresight.

Guinee, J., *et al.* (2002) *Life Cycle Assessment: An Operational Guide to the ISO Standards* (Part 2), The Hague: Ministry of Housing, spatial planning, and environment: http//:

Harrison, R. (1965) *Animal Machines,* London: Stuart.

Hart, H. L. A. (1961) *The Concept of Law,* Oxford: Clarendon Press.

Hawken, R. J., Beattie, C. W., and Schook, L. B. (1998) Resolving the genetics of resistance to infectious disease. *Scientific and Technical Review – International Office of Epizootics*, 17: 17–25.

Hobbes, T. (1651) *Leviathan:* www.gutenberg.org/ebooks/3207.

Hoekstra, A. Y. (2010) The water footprint of animal products. In J. D'Silva and J. Webster (eds), *The Meat Crisis* (pp. 22–33), London: Earthscan.

Hu, F. B., Manson, J. E., and Willett, W. C. (2001) Types of dietary fat and risk of coronary heart disease: a critical review. *Journal of American College Nutrition,* 20: 5–19.

Hulebak, K. L., and Schlosser, W. (2002) Hazard analysis and critical control point (HACCP) history and conceptual overview. *Risk Analysis,* 547–52.

Johnson, L. A. (1995) Sex preselection by flow cytometric separation of X and Y chromosome-bearing sperm based on DNA difference: a review. *Reproductive Fertility Developments,* 7: 893–903.

Jones, M. B., and Donnelly, A. (2004) Carbon sequestration in temperate grassland ecosystems and the influence of management, climate and elevated CO_2. *New Phytologist,* 164: 423–39.

Kaminski, S., Cielinska, A., and Kostyra, E. (2007) Polymorphism of bovine beta-casein and its potential effect on human health. *Journal of Applied Genetics,* 48: 189–98.

Kendrick, K. M. (1998) Intelligent perception. *Applied Animal Behaviour Science,* 57: 213–31.

Keys, A. (1980) *Seven Countries: a Multivariate Analysis of Death and Coronary Heart Disease*, Oxford: CAB Direct..

Kirkwood, J., and Webster, A. J. F. (1984) Energy budget strategies for mammals and birds. *Animal Production* 38, 147–56.

Knowles, T. G., Kestin, S. C., and Haslam, S. M. (2008) Leg disorders in broiler chickens, risk factors and prevention. *PLoS One,* 3(2): e1545.

Krecek, R. C., and Waller, P. J. (2006) Towards the implementation of the 'basket of options' approach to helminth parasite control of livestock: Emphasis on the tropics/subtropics. *Veterinary Parasitology,* 139: 270–82.

Land Degradation Assessment in Drylands (2006) Desertification and the international policy initiative, Joint International Conference Communication, 17–19 Dec., Algiers.

Maga, E. A., and Murray, J. D. (2010) Welfare applications of genetically engineered animals for use in agriculture. *Journal of Animal Science,* 288: 1588–91.

Mairi, J. (2007) The political economy of a productivist agriculture: New Zealand dairy discourses. *Food Policy,* 32: 266–79.

McInerney, J. (2012) Ethics and the economics of animal care. In C. M. Wathes (ed.), *First International Conference on Veterinary and Animal Ethics* (in press).

Mepham, B. (1996) Ethical analysis of food biotechnologies: an evaluative framework. In B. Mepham (ed.), *Food Ethics* (pp. 101–19), London: Routledge.

Moore, K., and Thatcher, W. W. (2006) Major advances associated with reproduction in dairy cattle. *Journal of Dairy Science,* 89: 1254–66.

National Research Council (USA) *Nutrient Requirements of Farm Animals* (multiple publications).

Nicol, C. (2011) Behaviour as an indicator of animal welfare. In UFAW, *Management and Welfare of Farm Animals,* 6th edn (ed. John Webster), ch. 2, Oxford: Wiley-Blackwell.

Nissani, M. (2004) Theory of mind and insight in chimpanzees, elephants, and other animals? *Comparative Vertebrate Cognition: Developments in Primatology: Progress and Prospects*, 4: 227–61.

OIE (Office International Epizootique) (2011) *Terrestrial Animals Health Code,* 20th edn; www.oie.int/international-standard-setting/terrestrial-code/access-online.

Packer, M. (2009) Algal capture of carbon dioxide: biomass generation as a tool for greenhouse gas mitigation. *Energy Policy,* 37: 3428–4339.

Pastorete, P.-P., Thiry, E., Brochier, B., Schwers, A., Thomas, I., and Dubuisson, J. (1998) Diseases of wild animals transmissible to domestic animals. *Scientific and Technical Review – International Office of Epizootics*, 7: 705–36.

Pelletier, N. (2008) Environmental performance in the US broiler poultry sector: life cycle energy use and greenhouse gas, ozone depleting, acidifying and eutrophying emissions. *Agricultural Systems,* 98: 67–73.

Pelletier, N., Lammers, P., Stender, D., and Pirog, R. (2010a) Life cycle assessment of high and low-profitability commodity and deep-bedded nache swine production systems in the upper Midwestern United States. *Agricultural Systems,* 103: 599–608.

Pelletier, N., Pirog, R., and Rasmussen, R. (2010b) Comparative life cycle environmental impact of three beef production strategies in the upper Midwestern United States. *Agricultural Systems,* 103: 380–9.

Potočnik, J. (2010) Can the CAP bring considerable benefits to our environment? 3rd Forum for the Future of Agriculture, The Economics and Politics of Food Security vs. Climate Change, Brussels, 16 March.

Pretty, J. (2008) Agricultural sustainability: concepts, principles and evidence. *Philosophical Transactions of the Royal Society B, Biological Sciences,* 363: 447–65.

Private Eye (2002) *Not the Foot and Mouth Report,* www.warmwell.com/footmoutheye.html.

Pryce, J. E., Royal, M. D., Garnsworthy, P. C., and Mao, I. L. (2004) Fertility in the high producing dairy cow. *Livestock Production Science,* 86: 125–33.

Pryce, J. E., Veerkamp, R. F., Thompson, R., Hill, W. G., and Simm, G. (1997) Genetic aspects of common health disorders and measures of fertility in Holstein Friesian dairy cattle. *Animal Science,* 65: 353–60.

Pursel, V. G., Bolt, D. J., Miller, K. F., Pinkert, C. A., Hammer, R. E., Palmiter, R. D., Brinster, R. L. (1990) Expression and performance in transgenic pigs. *Journal of Reproduction and Fertility,* 40: 235–45.

Radford, M. (2001) *Animal Welfare Law in Britain: Regulation and Responsibility,* Oxford: Oxford University Press.

Ramachandran Nair, P. K., Nair, V. D., Mohan Kumar, B., and Showalter, J. M. (2010) Carbon sequestration in agroforestry systems. *Advances in Agronomy,* 108: 237–307.

Roothaert, R. L., Ssalango, S., and Fulgensio, J. (2011) The Rakai Chicken Project: an approach that has improved fortunes for Ugandan farmers. *International Journal of Agricultural Sustainability,* 9: 222–31.

Royal Society (2009) *Reaping the Benefits: Science and the Sustainable Intensification of Global Agriculture*, London: Royal Society, www.royalsociety.org.

Russel, J. B. (1997) Mechanisms of action of ionophores. *Cornell Nutrition Conference for Feed Manufacturers,* 59: 88–92.

Schneider, M. J., Tait, R. G. Jr, Busby, W. D., and Reecy, J. M. (2009) An evaluation of bovine respiratory disease complex in feedlot cattle: impact on performance and carcass traits using treatment records and lung lesion scores. *Journal of Animal Science,* 87(5): 1821–7.

Scientific Advisory Committee on Nutrition (2003) *Salt and Health,* London: HMSO.

Sharvelle, S., and Loetscher, L. (2010) Anaerobic digestion of animal waste in Colorado. *Colorado State University Extension Document,* 1.227, Fort Collins: Colorado State University Publications.

Smith, A. (1776) *The Wealth of Nations,* London: Penguin Classics, 1990 edn.

Soil Association (n.d.) *Organic standards,* www.soilassocation.org.

Soller, M., and Andersen, L. (1998) Genomic approaches to the improvement of disease resistance in farm animals. *Scientific and Technical Review – International Office of Epizootics*, 17: 325–45.

Soussana, J. F., Allec, T., and Blanfort, V. (2010) Mitigating the greenhouse gas balance of ruminant production systems through carbon sequestration in grassland. *Animal,* 4(3): 334–40.

Speedy, A. W. (2003) Global production and consumption of animal-source foods. *Journal of Nutrition,* 133: 4048S–4053S.

Streiffer, R., and Ortiz, S. G. (2010) Animals in research: enviropigs. *Life Science Ethics 2010,* 3: 405–22.

Thompson, A. K., Shaw, D. I., Minihane, A. M., and Williams, C. M. (2008) Trans-fatty acids and cancer: the evidence reviewed. *Nutrition Research Reviews,* 21: 174–88.

Treguer, P., and Pondavin, P. (2000) Global change: silica control of carbon dioxide. *Nature,* 406: 358–9.

Turner, J. (2010) *Animal Breeding, Welfare and Society,* London: Earthscan Publications.

Universities Federation for Animal Welfare (UFAW) (2010) *The Management and Welfare of Farm Animals,* ed. J. Webster, Oxford: Wiley Blackwell.

Universities Federation for Animal Welfare (2011) *The Care and Management of Farm Animals,* 6th edn (ed. J. Webster), Oxford: Wiley-Blackwell.

US Department of Agriculture (USDA) *Food Composition,* PLACE: Food and Nutrition Center, Nutrient Data Laboratory; http://fnic.nal.usda.gov.

van Dijk, L., Pritchard, J., Pradhan, S. K., and Wells, K. (2011) *Sharing the Load,* Rugby: Practical Action Publishing, www.practicalactionpublishing.org.

Wall, R. J., A. M. Powell, *et al*. (2005) Genetically enhanced cows resist intramammary Staphylococcus aureus infection. *Nature Biotechnology,* 23: 445–51.

Ward, A. I., Tohurst, B. A., and Delahay, R. J. (2006) Farm husbandry and the risks of disease transmission between wild and domestic mammals: a brief review focusing on bovine tuberculosis in badgers and cattle. *Animal Science,* 82: 7676–7773.

Webster, A. J. F. (1986) Factors affecting the body composition of growing and adult animals. *Proceedings of the Nutrition Society,* 45: 45–53.

Webster, A. J. F. (1989) Bioenergetics, bioengineering and growth. *Animal Production,* 48: 249–69.

Webster A. J. F. (1992) Energy expenditure, studies with animals. In E. M. Widdowson and J. C. Mathers (eds), *The Contribution of Nutrition to Human and Animal Health* (pp. 23–32), Cambridge: Cambridge University Press.

Webster, A. J. F. (1993) *Understanding the Dairy Cow*, 2nd edn, London: Blackwell Scientific Publications.

Webster, A. J. F. (2009) The virtuous bicycle: a delivery vehicle for improved farm animal welfare. *Animal Welfare,* 18: 141–8.

Webster, A. J. F. (2012) Critical control points in the delivery of animal welfare. *Animal Welfare* 21(supp. 1), 117–124.

Webster, J. (1984) *Calf Husbandry, Health and Welfare,* London: Collins.

Webster, J. (1994) *Animal Welfare: A Cool Eye towards Eden,* Oxford: Blackwell Science.

Webster, J. (2005) *Animal Welfare: Limping towards Eden,* Oxford: Wiley-Blackwell.

Webster, J. (2011a) Anthropomorphism and zoomorphism: fruitful fallacies. *Animal Welfare,* 20: 29–36.

Webster, J. (2011b) Food from the dairy: husbandry regained? In J. D'Silva and J. Webster (eds), *The Meat Crisis,* London: Earthscan.

Webster, J., Bollen, P., Grimm, H., and Jennings, M. (2010) Ethical implications of using the minipig in regulatory toxicology studies. *Journal of Pharmacological and Toxicological Methods,* 62: 160–6.

Welfare Quality (2009) *Assessment Protocols for Cattle, Pigs and Poultry,* Lelystad, the Netherlands: Welfare Quality consortium.

Wemelsfelder, F., Hunter, E. A., Mendl, M. T., and Lawrence, A. B. (2001) Assessing the 'whole animal': a free-choice profiling approach. *Animal Behaviour,* 62: 209–20.

Whay, H. R., Main, D. C. J., Green, L. E., Heaven, G., Howell, H., Morgan, M., Pearson, A., and Webster, A. J. F. (2007) Assessment of the behaviour and welfare of laying hens on free-range units. *Veterinary Record*, 161: 119–28.

Whitehead, C. C. (2004) Overview of bone biology in the egg-laying hen. *Poultry Science,* 83: 193–9.

Winters, M. (2007) Genetics: New Productive Lifetime Index (*PLI*) will improve dairy cow health and fitness. *Livestock Production Science,* 12: 53–5.

Wirsenius, S., Azar, C., and Berndes, G. (2010) How much land is needed for global food production under scenarios of dietary changes and livestock productivity increases in 2030? *Agricultural Systems,* 103: 621–38.

World Cancer Research Fund (2007) *Food, Nutrition, Physical Activity and the Prevention of Cancer: A Global Perspective,* Washington, DC: American Institute for Cancer Research.

World Health Organization (2003) *Diet, Nutrition and the Prevention of Chronic Diseases,* Geneva: WHO.

Index